# GEOLOGY

## OF NEWFOUNDLAND

Touring through time at 48 scenic sites

Martha Hickman Hild

# GEOLOGY
## OF NEWFOUNDLAND
Touring through time at 48 scenic sites

Martha Hickman Hild

Library and Archives Canada Cataloguing in Publication

© 2012 Hild, Martha Hickman
   Geology of Newfoundland: field guide / Martha Hickman Hild.

Includes bibliographical references and index.
ISBN 978-1-927099-07-0

1. Geology--Newfoundland and Labrador--Newfoundland, Island of--Guidebooks.
2. Newfoundland, Island of (N.L.)--Guidebooks.  I. Title.

QE199.H54 2012       557.18       C2012-901270-X

Published by Boulder Publications
Portugal Cove-St. Philip's, Newfoundland and Labrador
www.boulderpublications.ca

Editor: Stephanie Porter
Copy editor: Iona Bulgin
Cover design and page layout: Sarah Hansen

Front cover: Green Point (top) and Tilting (bottom)
Back cover: Manuels River

Printed in China

**Newfoundland Labrador** We acknowledge the financial support of the Government of Newfoundland and Labrador through the Department of Tourism, Culture and Recreation.

**Canadä** We acknowledge financial support for our publishing program by the Government of Canada and the Department of Canadian Heritage through the Canada Book Fund.

## Dedication

To the memory of Dr. Harold (Hank) Williams (1934-2010),
who inspired generations of geologists with a lifetime of energetic
commitment to field-based science.

Hank believed "the truth lies in the rocks."

# Table of Contents

# Author's Preface

G eology of Newfoundland will help you find, understand, and enjoy the rocks at a variety of scenic, accessible sites on the island: The book describes 48 sites of primary interest as well as additional, related outcrops – more than 90 locations in all.

This project has its origins in a long-held tradition – the geological field trip. The late Hank Williams's week-long trans-island field trips were a particular inspiration. Several of the sites in this book were among those he used to bring the history of the Appalachian mountains alive for students and colleagues. Because the geology of Newfoundland is unique and varied, geologists come from all over the world to visit significant outcrops. The geological community's appetite for such activities has led to the publication of dozens of field guides. Why create another one?

Most field guides have been written by professional geologists for use by colleagues. Geologists don't mind teetering on the edge of a ditch to look at rocks while trucks pound past on the highway. They take high-clearance vehicles into the field and don't mind bouncing along kilometres of narrow forestry road – then sloshing through a bog. Because of this, many of the well-established sites included in their field guides are, although informative, not suitable for a casual trip or family outing.

Some field guides lack detail because they are meant to supplement information provided by a qualified leader, who shows the way to the outcrop and points out features of interest. Others contain lengthy passages from scientific articles – difficult reading for non-specialists. Many are out of print, and many were published informally without maps or photos.

As I reviewed the existing field guides for Newfoundland, I envisioned a different kind of book, one that makes it easy for anyone to find interesting rocks while experiencing the island's natural beauty. I also saw the need for a field guide that tells the story of how Newfoundland formed as part of the Appalachian mountains.

The first challenge in creating such a book lay in the selection of sites. I consulted several geologists and pored over previously published field guides. My search focused on locations that are scenic, accessible, and representative of the island's many geographic regions, as well as its geological past. I have visited each of the sites mentioned in this book as well as many others, selecting only those that I found pleasant and engaging.

Another challenge was the book's design. I found it difficult to plan my trip using existing field guides: Would the site suit my interests? How much time would I need to invest in getting to the outcrop? I did some head-scratching in the field, too, and on my travels I encountered fellow geo-tourists in the same quandary. Where exactly is the outcrop? How do the rocks fit into a sequence of geological events? Existing field guides were not always clear.

My solution was to design a series of structured, 4-page site descriptions (6 pages for sites with multiple points of interest) that provide all the information you need to have a good time looking at rocks. Complementing the site descriptions are other resources to help you explore topics further.

If you use this book to visit outcrops in the field, I hope you'll find, as I did, that tracking down Newfoundland's geological past can provide wonderful recreation. The rocks tell a fascinating story and are well exposed in the province's parks, reserves, and tracts of public land. They are intertwined with the history of Newfoundland's people as well as the future of its economy.

Many communities on the island have made significant efforts to welcome visitors with related attractions, trails, boardwalks, or stairways that make it easy and fun to reach sites of geological importance as part of a larger nature- or history-oriented outing. Go on – get out there and explore this place we affectionately call The Rock.

# Acknowledgements

This field guide would not have been possible without the help of many people. Greg Dunning of Memorial University allowed me to audit his course on the geology of Newfoundland and participate in the course's trans-island field trip in the spring of 2011. Greg suggested several of the sites included here and patiently answered many questions about Newfoundland's rocks and their history.

Mark Wilson brought me up to speed on the world of GIS data and software, helped me refine my approach to field research for the book, and has been an invaluable sounding board throughout the project. He and Pauline Honarvar travelled with me on the Great Northern peninsula; their supportive encouragement and their hospitality in Raleigh deserve special mention.

Others who replied to questions with helpful suggestions and information include Sean O'Brien and Greg Sparkes of the Newfoundland and Labrador Geological Survey; Rick Hiscott, Toby Rivers, and Joe Hodych of Memorial University; Howard Falcon-Lang of the University of London; Arden Bashforth of the Smithsonian Institution; Ed Landing of the New York State Museum; and Paul Myrow of Colorado College. For their patience and expertise, I offer my sincerest thanks. Unnamed but indispensable contributors include dozens of additional geologists who over the past four decades have endured the rigours of fieldwork in Newfoundland to unravel the complex geological history of this wonderful island.

The team at Boulder Publications has moved the project from proposal to reality with great patience, focus, and energy. I am grateful to Gavin Will for believing in the need for this book, and to Iona Bulgin, Sarah Hansen, Stephanie Porter, and Vanessa Stockley for the hard work and talent they brought to the project. The field guide also benefitted from generous investments of time by earth science professionals who reviewed the manuscript: Greg Dunning, Mark Wilson, and Sharon Deemer.

For cheering me on, I send out a big thank you to my dad, siblings, daughters, and especially to my husband, Dave – for surviving capably through my absences in the field and my reclusive writing habits, for keeping my old car roadworthy for 12,000 kilometres of field travel, and for drafting several of the figures.

# How to Use This Guide

## What's in the Guide

This book is organized as a journey through time and aims to include all the "provisions" you'll need to make the trip. Here's a brief summary.

### Sample Pages

Pages 4 and 5 provide a graphical key to the field guide's 48 site descriptions.

### Geology Basics

This section is for readers who want basic background information about geology. How do geologists measure and describe geologic time? What are the main kinds of rocks and how do they form? How do tectonic plates interact? How did the Appalachian mountains form? These topics are briefly reviewed at the front of the book.

### Resources

For the extra curious, following Geology Basics is a list of geology-themed museums and interpretive centres as well as print and online resources you can explore to learn more.

### Trip Planner

On pages 22 and 23 is a Trip Planner you can use to select sites and plan your itinerary based on what interests you, what parts of the island you'll be visiting, how far you want to hike, and other preferences. The Trip Planner lists the 48 sites covered in the book as well as several geology-related museums and interpretive centres.

### Zones at a Glance

The site descriptions are grouped into four sections, one for each of Newfoundland's tectonic zones – Humber, Dunnage, Gander, and Avalon – from west to east. Each section begins with a summary that includes a map of the zone, a list reviewing what you can experience at each site, and a brief account of geological events.

## Site Descriptions

The core of the book describes 48 sites of geological interest. The journey begins in the Humber zone, where Newfoundland's oldest rocks are found. The Humber zone is the only part of Newfoundland that has always been attached to the rest of North America, so it's a logical place to begin the story. The other zones are presented in the order in which they were joined to the continent: after Humber, then Dunnage, next Gander, and finally Avalon.

## Glossary

Words that may be unfamiliar are explained in the glossary at the back of the book.

## Index

An alphabetical index of place names includes all locations mentioned in the text. There are more than 90 entries in all.

# Legends

## Tectonic Zones

Throughout the book, specific colours are associated with each of Newfoundland's tectonic zones.

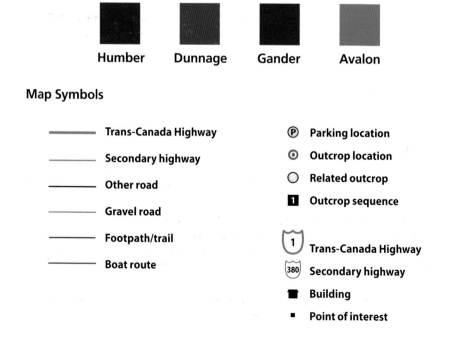

| Humber | Dunnage | Gander | Avalon |

## Map Symbols

——— Trans-Canada Highway

——— Secondary highway

——— Other road

——— Gravel road

——— Footpath/trail

——— Boat route

Ⓟ Parking location

◎ Outcrop location

○ Related outcrop

**1** Outcrop sequence

⬡ 1 Trans-Canada Highway

⬡ 380 Secondary highway

◼ Building

▪ Point of interest

NOTE: All maps are oriented with North at the top (except those on pages 16 and 155).

## Icons

NOTE: Icons relevant to the site are printed in a dark, solid colour. Icons not applicable are printed in a lighter tone.

 **Walking distance** – This is the approximate total walking distance for the site, that is, the distance from the parking location to the outcrop and back.

 **Water level** – This outcrop is best viewed in low water conditions (low tide or low lake and river levels). Check tide tables or other sources when planning your visit. NOTE: This icon refers to normal tides and normal seasonal water-level fluctuations only. Storms or other unusual conditions affecting water levels may pose safety and access issues even for sites not marked with this icon.

 **Seasonal** – Access to this outcrop is seasonal due to park closure. NOTE: This icon is not intended to indicate the effect of weather, road conditions, or other seasonal factors on site access.

 **Park** – This site is located within a national, provincial, or municipal park or reserve. It is your responsibility to be aware of any park rules and policies affecting your visit.

 **Cost** – There is a fee to visit this site and/or the park in which it is located.

## Also Note

- For each site, a map number and name are listed under the heading 1:50,000 Map. These identify the National Topographic System of Canada (NTS) map on which the site appears.

- All latitude-longitude readings in the field guide are given in the same format, for example, N49.74376, W57.77707. The letter prefix refers to the hemisphere (northern latitude and western longitude) and should be included when entering the coordinates into a GPS device or internet mapping utility such as Bing Maps, Google Maps, or Google Earth.

- In many of the photos, a metric scale is visible. The scale is 10 centimetres long; its smallest subdivisions are 1 centimetre square.

# Sample Pages

**Site location** in easy-to-reference heading, colour-coded by zone.

**Local scenery** to help you recognize the site when you arrive.

Non-technical **description** of significance and features at the site.

---

2. FLOWER'S COVE

Thrombolites emerge as the tide falls in Flower's Cove.

## Tropical Paradise
### Thrombolite Mounds at Flower's Cove

When the Iapetus ocean opened and began to widen, western Newfoundland was on the edge of a new continent, Laurentia. Its stable continental margin was free from crustal upheaval or volcanic activity. Plate tectonic movements had positioned the region near the equator – expanses of warm ocean supported many new life forms.

Thrombolites are limestone mounds formed by colonies of ocean-dwelling microbes. They were common at this point in Earth history, first appearing early in the Cambrian period. By the end of the Cambrian period, they had reached the peak of their success.

Throughout the Ordovician period their abundance declined, and they all but disappeared as other life forms took over the "mound business," building structures better able to resist burrowing and predation by an evolving army of new marine life (for example, see site 4).

At Flower's Cove many of the fossil thrombolites are completely exposed at low tide. But when they were alive long ago, they were probably in slightly deeper water, so that at low tide the water just covered their broad, flat tops.

34

**1**

---

### On the Outcrop

Cambrian thrombolite mounds merged into a large group.

**Outcrop Location: N51.29350, W56.74253**

Emerging like a tray of gigantic macaroons from the shallow water of the cove, the thrombolites of Flower's Cove form mounds 40 to 200 centimetres or more in diameter.

Many kinds of marine life create mound-like structures (for example, the corals of our modern oceans). As life evolved, organisms developed the ability to form exoskeletons – hard, protective shells – with distinctive shapes for each individual species. The Cambrian thrombolites represent a more primitive strategy and were built up of small clumps of calcified microbe colonies.

Superficially, thrombolites resemble stromatolites (see site 45) in that both form mounds, and both were common in the Cambrian period. But stromatolites are built up of distinct layers, often visible as fine concentric lines of sediment within the mound. As you'll see, there is no such layering within the mounds at Flower's Cove.

Like modern coral reefs, thrombolites provided a habitat for many other organisms living within and among the mounds. The most obvious evidence for this is the numerous burrows that puncture the mounds.

36

1000    900    800    700    600

**3**

**Image** of the outcrop to help you find and interpret the rocks.

**GPS waypoint** for the outcrop location – nature's geocache. Provided in decimal degrees so you can easily enter it into your GPS device or web browser.

**Description** of the outcrop emphasizing what to look for.

4

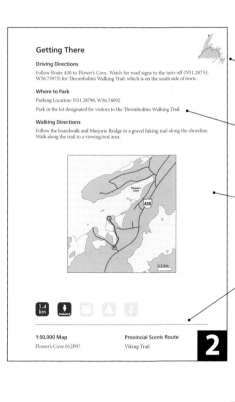

## Getting There

**Driving Directions**

Follow Route 430 to Flower's Cove. Watch for road signs to the turn-off (N51.28733, W56.73973) for Thrombolites Walking Trail, which is on the south side of town.

**Where to Park**

Parking Location: N51.28798, W56.74092

Park in the lot designated for visitors to the Thrombolites Walking Trail.

**Walking Directions**

Follow the boardwalk and Marjorie Bridge to a gravel hiking trail along the shoreline. Walk along the trail to a viewing/rest area.

1:50,000 Map
Flower's Cove 012P07

Provincial Scenic Route
Viking Trail

**2**

**Thumbnail map** showing site location.

Detailed written **directions** for driving, parking (including GPS waypoint), and walking.

Road map showing the **route** to the outcrop from the nearest highway (for Legend, see page 2).

At-a-glance **information** to help you plan your trip: key site characteristics plus topographic map and tourism route details.

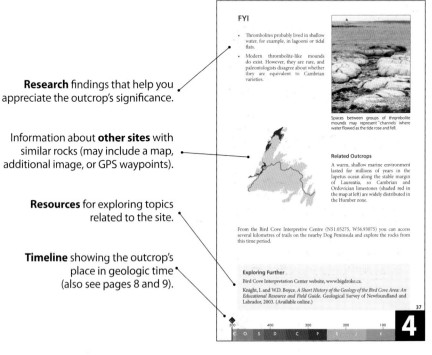

## FYI

- Thrombolites probably lived in shallow water, for example, in lagoons or tidal flats.
- Modern thrombolite-like mounds do exist. However, they are rare, and paleontologists disagree about whether they are equivalent to Cambrian varieties.

Spaces between groups of thrombolite mounds may represent "channels" where water flowed as the tide rose and fell.

### Related Outcrops

A warm, shallow marine environment lasted for millions of years in the Iapetus ocean along the stable margin of Laurentia, so Cambrian and Ordovician limestones (shaded red in the map at left) are widely distributed in the Humber zone.

From the Bird Cove Interpretive Centre (N51.05275, W56.93075) you can access several kilometres of trails on the nearby Dog Peninsula and explore the rocks from this time period.

### Exploring Further

Bird Cove Interpretation Center website, www.bigdroke.ca.

Knight, I. and W.D. Boyce. *A Short History of the Geology of the Bird Cove Area: An Educational Resource and Field Guide.* Geological Survey of Newfoundland and Labrador, 2003. (Available online.)

37

**4**

**Research** findings that help you appreciate the outcrop's significance.

Information about **other sites** with similar rocks (may include a map, additional image, or GPS waypoints).

**Resources** for exploring topics related to the site.

**Timeline** showing the outcrop's place in geologic time (also see pages 8 and 9).

# Exploring the Sites

## Safety

You can't experience geology directly without going outdoors. All the sites in this book have been visited safely by many people. These are places that residents, tourists, geologists, students, and others go to enjoy the beauty of the province and examine interesting rocks. However, weather, tides, and other conditions can render any site hazardous temporarily. Only you can decide whether it is safe to visit a specific site on a specific day, and whether your state of preparedness is appropriate for the conditions.

## Navigating

Each site description provides information to help you find the outcrop easily. Road maps and written directions indicate the route from the nearest highway to the outcrop. There is also information to help you find the site on topographic or tourism maps.

Field readings of GPS latitude and longitude are provided for parking and outcrop locations at all sites. The readings reference World Geodetic System datum WGS84. If you enter them into your own GPS device, web browser, or other application, keep in mind that results may vary depending on your GPS device settings, map projections, application preferences, and other factors. The readings are not intended, and should never be used, as a substitute for attentive real-world navigation.

The parking locations given are example locations where parking was possible at the time of publication. Please exercise common sense and courtesy at all times when parking your vehicle, based on the conditions you encounter. For most sites, the listed outcrop location marks the position of the rock described in the text. Occasionally, though, the outcrop location is a vantage point from which significant features can be viewed.

Written directions to the sites include words and phrases such as "northward," "to the east," or "southwest." These are used only in a general sense to clarify correct choices at forks, intersections, and other turning points. They are not intended as precise orienteering instructions. Driving and hiking distances are approximate and are only provided as an aid to finding the sites and outcrops.

## Preserving Outcrops

The geology of Newfoundland is unique. If you use this field guide to visit sites, please apply the principles of Leave No Trace Canada (www.leavenotrace.ca) to preserve Newfoundland's natural beauty and geological treasures for others to enjoy. Each of the selected sites offers a clear view of well-exposed rock. Don't hammer outcrops or remove material – photographs are the best way to capture your experiences.

## Protected Sites

It's always preferable to leave sites undisturbed for others to enjoy, but you should also be aware that certain areas of the province are protected by law from any hammering or rock collecting. In addition, it is against the law to remove fossil material from many sites in Newfoundland without a permit. The second page of each unit (Getting There) identifies sites protected at the time of publication. However, protected status of sites is subject to change. The Newfoundland and Labrador Department of Natural Resources is the best source for current information on protected natural areas and sites.

# Geology Basics

## Geologic Time

### Superposition and Cross-Cutting

The features of an outcrop can reveal the relative ages of geologic events, that is, the order in which they occurred. The principles of superposition and cross-cutting allow you to reconstruct a sequence of events even though they happened long ago.

**Superposition**. Sedimentary rocks are deposited in flat layers, with younger layers on top of older layers.

**Cross-cutting**. Any feature (intrusion, fault, erosion) that cuts across or truncates other features is younger than all the features it cuts across.

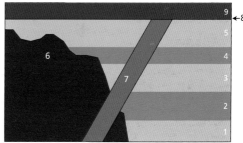

A simplified rock outcrop showing the order in which features formed: 1-5, sedimentary layers; 6, igneous intrusion; 7, cross-cutting dyke; 8, erosion surface; 9, a younger sedimentary layer.

### Fossil Species and Radiometric Data

Geologists use two types of evidence – fossil species and radiometric data – to tell when rocks formed and geologic events happened.

**Fossil species**. As life evolved in the geologic past, new species appeared, existed for a time, and became extinct. Based on close study of the appearance, distribution, and extinction of fossil species preserved in rock, the international geological community has agreed on standards that define which fossils belong to which geologic period (see sites 3 and 43). This allows a rock layer to be assigned an age based on the fossils it contains.

**Radiometric data**. Some rocks and minerals contain small amounts of radioactive atoms. Radioactive decay converts an unstable atom (called the parent isotope) into a stable atom (called the daughter isotope). For example, radioactive atoms of uranium decay to form stable atoms of lead. By measuring the amount of parent and daughter isotopes very precisely, geologists can calculate the age of the rock using the half-life of the radioactivity.

# Proterozoic Eon

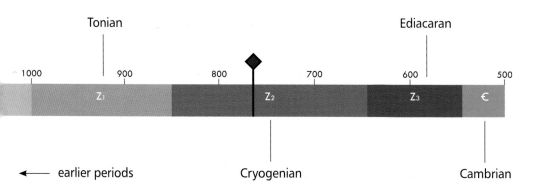

*Late Proterozoic Era*

Tonian

Ediacaran

1000    900    800    700    600    500

Z₁    Z₂    Z₃    €

◄── earlier periods

Cryogenian

Cambrian

## The Book's Timeline

On the third and fourth pages of each site description is a ribbon of colour like the one shown above. It represents part of the geologic time scale – about one quarter of Earth history.

The numbers on the timeline mark off intervals of 100 million years before the present day; the colours mark the boundaries of geologic periods and eras; and the letters are abbreviations for the names of the periods (except Cenozoic, which is an era, not a period).

Eras are groups of geologic periods. Eras are shown on the timeline by colours: shades of brown for the Neoproterozoic (or late Proterozoic) era; shades of blue for the Paleozoic era; shades of green for the Mesozoic era; and yellow for the Cenozoic era.

Each site description includes a marker on the timeline showing the age of the rock or event of interest at that site. In this example the marker is placed at 765 million years ago.

## Exploring Further

Edwards, Lucy and John Pojeta, Jr. *Fossils, Rocks, and Time* (Online Edition). US Geological Survey website, pubs.usgs.gov/gip/fossils.

# Phanerozoic Eon

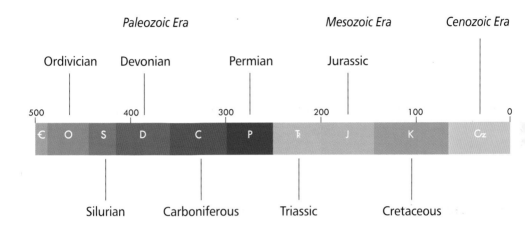

*Paleozoic Era*          *Mesozoic Era*     *Cenozoic Era*

Ordivician    Devonian          Permian        Jurassic

500            400          300            200           100              0

€    O    S    D        C        P        T̵        J         K            Cz

Silurian      Carboniferous     Triassic        Cretaceous

## Four Eons of Geologic Time

The book's timeline shows only part of the history of the Earth – our planet formed about 4,550 million years ago. Geologists divide geologic time into eons, which are subdivided into eras and periods. Here's a quick summary of what happened during each eon:

Hadean eon (4,550 to 4,000 million years ago). Very few rocks survive from the Hadean eon. The Earth's crust was hot, mobile, and bombarded by meteors.

Archean eon (4,000 to 2,500 million years ago). Tectonic plates and early continents formed; primitive single-celled micro-organisms appeared in the sea.

Proterozoic eon (2,500 to 542 million years ago). Areas of stable continental crust grew as plate tectonics caused a series of continental collisions, creating mountain belts. Oxygen levels increased in the air and sea, allowing early complex life forms to evolve.

Phanerozoic eon (542 million years ago to the present). Proterozoic continents assembled to form Pangaea. Pangaea broke up as present-day continents moved to their current positions. Animals and plants evolved to inhabit both sea and land.

# Rock Types

## Identifying Rocks

"What kind is it?"

That's the first question that comes to mind when you see a rock in the field. Correctly identifying the type of rock you are looking at is the first step in reading the rock's story about the geological past.

Newfoundland provides easy access to an extraordinary variety of rock types, formed in a wide range of conditions – it's a natural showcase of Earth materials.

There are plenty of general guides to rock identification in print and online. This brief review uses examples from the sites described in this book.

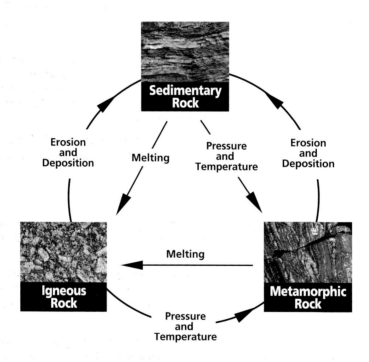

The rock cycle shows how igneous, sedimentary, and metamorphic rocks can be transformed and recycled into other rock types when exposed to new conditions. Changing conditions can be caused by uplift, burial, the movement of fluids in the crust, and other events as the Earth's tectonic plates interact.

## Igneous Rocks

Igneous rocks form when molten rock (magma) rises through the crust, cools, and crystallizes. Igneous rocks are classified based on the minerals they contain, and whether they are intrusive (crystallized slowly underground) or extrusive (erupted onto the surface and crystallized quickly).

At one end of the spectrum, granites contain mostly light-coloured minerals (quartz and feldspar), often with small amounts of dark minerals such as biotite or amphibole (sites 1, 22, 30, 32, 34, 36, 40, and 48). True granites contain potassium feldspar, plagioclase, and quartz. However, the term "granitic" is used informally to refer to a wide variety of igneous rocks dominated by any combination of light-coloured minerals.

Intermediate rocks include trondhjemite (site 14) and tonalite (site 23) as well as granodiorite and diorite. They contain a mixture of light and dark minerals; the ones richer in quartz are often grouped with true granites using the informal term "granitic."

Mafic rocks contain mostly dark minerals such as pyroxene, but with small amounts of feldspar (sites 1, 9, 10, 13, 15, 16, 23, 35, and 47). Gabbro, diabase, and basalt all have a similar mixture of minerals but differ in the size of the mineral grains – easily visible in gabbro to microscopic in basalt.

Ultramafic rocks are found in the Earth's mantle but rarely appear at the surface. They are made almost exclusively of dark, dense minerals such as pyroxene and olivine. From large tracts of peridotite (site 11) to smaller amounts of websterite (site 23), as well as numerous occurrences of altered ultramafic rock (sites 18 and 27), Newfoundland offers plenty of opportunities to view this unusual category.

Another notable, though rare, rock type found in Newfoundland is lamprophyre (site 25). Mica-rich and very high in potassium, it doesn't fit into the commonly used classification schemes.

## Sedimentary Rocks

Sedimentary rocks form in layers. The layers originate as loose sediment, usually deposited underwater. They harden into rock as they become deeply buried by more sediment.

Some sedimentary rocks are made of broken-up, weathered fragments eroded from older rocks. The particle size of the sediment determines the rock type: Tiny clay particles form shale; sand forms sandstone (or, if it's very fine, siltstone); and gravel and larger fragments form conglomerate.

Other sedimentary rocks form chemically when minerals precipitate directly from water. (Sometimes the process is aided by biological activity.) Carbonate rocks like limestone and dolostone form in this way.

Newfoundland has wonderful examples of these common sedimentary rocks. Limestone is especially abundant in the Humber zone (sites 2 and 4), in some places alternating with layers of shale (site 3). Siltstone and sandstone can be found in every zone. Some formed in the ocean (sites 26 and 46), often as turbidites (sites 5, 6, 17, and 38). Turbidite layers form as a slurry of sandy sediment swirls down a steep underwater slope.

Other siltstones and sandstones in Newfoundland were deposited by rivers flowing from regions of tectonic uplift (sites 24, 41, and 42). Conglomerates formed on steep slopes in the ocean (sites 5 and 21) or where rivers flowed rapidly (sites 12 and 41).

Some of Newfoundland's less common sedimentary rocks have challenged the geologists trying to classify them. Chaotic sedimentary block-in-matrix melanges formed as rock fragments of many kinds and sizes fell into deep accumulations of ocean mud (sites 8 and 20). Debris flows, like underwater landslides, dislodged existing sediment and deposited a well-mixed, thick slurry at the bottom of a steep slope (sites 5 and 37). Rare conditions of ocean chemistry during rapid climate change led to the precipitation of an unusual post-glacial carbonate (site 37).

## Metamorphic Rocks

Metamorphic rocks form when existing rocks of any type recrystallize due to changes in temperature and/or pressure – for example, when a region of the Earth's crust is buried under a colliding tectonic plate. Metamorphic rocks are classified based on the intensity, or grade, of the metamorphism. The different grades are defined based on the presence of specific metamorphic minerals.

Much of Newfoundland has never been deeply buried, so many of its rocks are only slightly metamorphosed. Sometimes the only sign of metamorphism is a greenish colour in the rock caused by the presence of the low-grade metamorphic mineral chlorite (sites 13 and 26).

A few areas of the island have been very deep in the Earth – 25 kilometres or more. At that depth, minerals such as garnet and kyanite form (site 29). Certain combinations of minerals may even melt at this depth (site 28).

During tectonic plate interactions, rocks can also be folded (sites 7 and 28) as crustal blocks move toward one another and compress the terrain. When this happens, flaky minerals like mica may become aligned, creating rock types that split apart easily, such as slate, schist, or gneiss (sites 28, 29, and 44).

Some metamorphic rocks show signs of intense deformation. Newfoundland's many melanges are good examples of how movement between blocks of crust can be accommodated in a narrow zone of easily deformed rock (sites 6, 7, 8, 9, and 20). At the tectonic boundaries between crustal blocks, relative movement caused shear and fault zones (sites 19, 31, and 33) that are also intensely deformed.

## Exploring Further

Bishop, Arthur C., Alan R. Wooley, and William R. Hamilton. *Guide to Minerals, Rocks & Fossils*. Firefly Books Ltd., 2005.

# Plate Tectonics

The story of Newfoundland's geological past is in large part a story of plate tectonics. In this section you'll find some of the ideas and terms geologists use when thinking and writing about tectonic plates.

## Tectonic Plates

Surrounding the Earth's hot, molten outer core is the mantle. Solid but capable of flowing, the mantle carries heat toward the surface of the planet by circulating slowly. Hot regions of the mantle rise toward the Earth's surface and cooler regions of the mantle sink back toward the core in a process of mantle convection.

The Earth's crust (including all the continents and ocean basins) plus the cooler, upper part of the mantle form a thin outer layer called the lithosphere. Since at least 2,500 million years ago, the lithosphere has been rigid enough to form well-defined tectonic plates. The continents and ocean basins are in a constant state of change, with most changes taking place along plate boundaries.

Because the mantle beneath them is convecting, tectonic plates move relative to one another at a rate of several centimetres per year. Plate motion causes earthquakes and volcanic activity along plate boundaries, making present-day boundaries easy to recognize.

## Plate Interactions

There are three types of plate motion: divergent, convergent, and transform. For plate motion occurring at the present day, distinctive patterns of earthquake and volcanic activity reveal the boundary type. But what about plate motions in the past? Igneous, sedimentary, and metamorphic rocks provide clues to plate interactions that took place millions of years ago, because plate interactions are what cause rocks to form and change. Each type of plate interaction leaves a "signature" of rock characteristics.

## Divergent Boundaries

At divergent plate boundaries, hot regions of the mantle well upward and then flow away from the plate boundary, moving the plates apart. The Earth's crust cracks open, causing shallow earthquakes. Molten rock rises through the cracks, creating intrusions below the surface and volcanic eruptions above.

Divergent boundaries may begin within an area of continental crust. Dyke swarms mark areas of crust that have been pulled apart in this way. As rifting continues, a steep-sided rift valley forms and deepens over time as the continental crust becomes thinner in the floor of the valley. Eventually the continental crust separates completely and a narrow zone of ocean crust develops.

If the plates continue to move apart, the ocean basin widens. This part of the process is called the rift-drift transition. Along the shores of the separated continents, the steep valley walls formed by rifting are replaced over time by stable continental margins known as passive margins. In Newfoundland, rift-drift transitions from both Laurentia (see site 1) and Gondwana (see site 46) are preserved.

Along a typical passive margin, the shallow continental shelf is followed seaward by the deeper-water continental slope, continental rise, and finally the ocean floor itself. Many of the sedimentary rocks in Newfoundland formed in a passive margin environment along the shores of Iapetus.

Mid-ocean ridges mark divergent plate boundaries within areas of ocean crust. To date, very little ocean crust formed at the Iapetus mid-ocean ridge has been found, presumably because it was all carried back into the Earth via subduction along convergent boundaries.

Long, narrow seaways sometimes form by rifting within a volcanic arc or between a volcanic arc and a nearby continent. The ocean crust formed in these back-arc basins has a characteristic chemical composition shared by some examples of ocean crust preserved in Newfoundland.

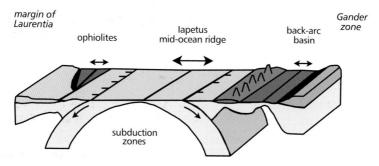

Rocks in Newfoundland formed in a plate tectonic setting that included divergent and convergent motions. Note the volcanic arc above the subduction zone at right.

## Convergent Boundaries

At convergent plate boundaries, cool regions of the mantle flow toward the plate boundary, then down into the Earth. The plates move toward one another. Convergent boundaries cause deep-seated, severe earthquakes and the Earth's most extreme topography in the form of ocean trenches and mountain ranges.

New convergent boundaries form within ocean crust, usually in older, cooler regions away from the spreading centre. The plate tears apart as one edge sinks, forming a subduction zone. Paradoxically, new ocean crust sometimes forms near a subduction zone. Examples of this are preserved in western Newfoundland in the form of ophiolites.

Once established, the subduction zone carries one of the two plates down into the mantle. A deep ocean trench forms along the plate boundary. As the ocean crust descends, it melts; the molten rock rises into the overriding oceanic plate, creating an island arc – a chain or cluster of islands above the subducting plate.

If there is continental crust along a convergent boundary, subduction usually carries the ocean crust below the continent because ocean crust is denser. When the ocean crust sinks deep enough to melt, a magmatic arc of igneous intrusions and volcanoes forms along the continental margin. Rarely, ocean crust is obducted – that is, pushed upward and onto the continent – instead of being subducted beneath it. This happened in western Newfoundland as a narrow seaway of Iapetus closed early in the history of the Appalachian orogeny.

If ocean crust and continental crust converge over millions of years, islands, microcontinents, and other landmasses in the ocean basin will collide with the continent in a process of crustal accretion. Because ocean crust is destroyed along convergent boundaries, the end result of plate convergence is continental collision. When two blocks of continental crust collide, one of the blocks may ride over the other, but continental crust cannot be subducted – it is too buoyant. Instead, continental crust in the collision zone becomes thicker. Lower regions of the thickened crust become hot and melt, forming granites, other intrusions, and metamorphic rocks.

## Transform and Complex Boundaries

At transform boundaries, tectonic plates slide past one another. Transform motion distorts the deeper, hotter regions of the Earth's crust along shear zones; earthquakes occur as shallower regions of the crust break along faults. Transform motion can break the crust in complicated patterns, forming deep basins between areas of uplift along faults.

## The Wilson Cycle

What is the outcome of all this plate motion during vast periods of geologic time? Ocean crust is constantly being created and destroyed, while continents slowly grow in size.

The theory of plate tectonics led to the realization that throughout Earth history a series of ocean basins opened and closed, causing periodic continental collisions of which mountain belts were the aftermath. This concept, now widely accepted, is known as the Wilson Cycle after its author, Canadian geophysicist J. Tuzo Wilson. The Wilson Cycle describes the "life cycle" of an ocean basin. Each cycle ends in continental collision, creating a new mountain range.

### The Wilson Cycle

| Stage | Event | Example |
|-------|-------|---------|
| 1 | Rifting | East African rift valley |
| 2 | Young ocean | Red Sea |
| 3 | Mature ocean | Atlantic Ocean |
| 4 | Declining ocean | Western Pacific Ocean |
| 5 | Terminal ocean | Mediterranean Sea |
| 6 | Ocean relict | Ophiolites in Himalayas |

## Exploring Further

Kious, W. Jacquelyne and Robert I. Tilling. *This Dynamic Earth: The Story of Plate Tectonics* (Online Edition). US Geological Survey website, pubs.usgs.gov/gip/dynamic. (An excellent summary with numerous illustrations, it can be viewed online or downloaded as a PDF.)

NASA Goddard Space Flight Center, Digital Tectonic Activity Map. US National Aeronautics and Space Administration website, denali.gsfc.nasa.gov/dtam.

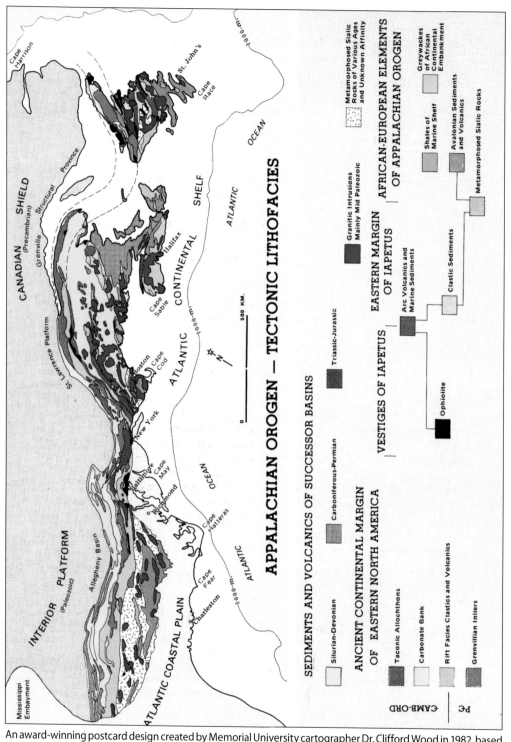

# APPALACHIAN OROGEN — TECTONIC LITHOFACIES

## SEDIMENTS AND VOLCANICS OF SUCCESSOR BASINS

- Silurian-Devonian
- Carboniferous-Permian
- Triassic-Jurassic

**ANCIENT CONTINENTAL MARGIN OF EASTERN NORTH AMERICA**

- Taconic Allochthons
- Carbonate Bank
- Rift Facies Clastics and Volcanics
- Grenvillian Inliers

**VESTIGES OF IAPETUS**

- Ophiolite
- Granitic Intrusions Mainly Mid Paleozoic

**EASTERN MARGIN OF IAPETUS**

- Arc Volcanics and Marine Sediments
- Clastic Sediments
- Metamorphosed Sialic Rocks
- Metamorphosed Sialic Rocks of Various Ages and Unknown Affinity

**AFRICAN-EUROPEAN ELEMENTS OF APPALACHIAN OROGEN**

- Shales of Marine Shelf
- Avalonian Sediments and Volcanics
- Greywackes of African Continental Embankment

CAMB-ORD / Pꞓ

### Map labels

CANADIAN SHIELD (Precambrian)

Grenville Structural Province

INTERIOR PLATFORM (Paleozoic)

Mississippi Embayment

St. Lawrence Platform

Allegheny Basin

ATLANTIC COASTAL PLAIN

ATLANTIC OCEAN

ATLANTIC CONTINENTAL SHELF

Cape Harrison

St. John's

Cape Race

Halifax

Cape Sable

Boston

Cape Cod

New York

Cape May

Baltimore

Richmond

Cape Hatteras

Cape Fear

Charleston

500 KM.

An award-winning postcard design created by Memorial University cartographer Dr. Clifford Wood in 1982, based on Dr. Hank Williams's detailed 1:1,000,000 scale, 3.4-metre-long map, "Tectonic Lithofacies of the Appalachian Orogen." The full-scale map was published by Memorial University in 1978. It was the first geologic map of an entire orogen, and the first map to apply the concepts of plate tectonics to the depiction of regional geology.

# The Appalachian Mountains

Newfoundland lies at the northeast end of the Appalachian mountain belt and provides a detailed, well-preserved, and easily accessed cross-section of the mountain belt. This section explains the Newfoundland origins of a unique geologic map of the Appalachians and outlines significant events in the history of the mountain belt.

## A Scientific Revolution

As recently as 1960, most geologists believed the Earth's crust was static and its continents fixed. They had no clear or plausible idea how mountain belts formed. In the absence of a unified theory that explained how the Earth functions, the science of geology focused on the description and classification of rocks, minerals, and fossils.

J. Tuzo Wilson helped develop the theory of plate tectonics, a new understanding of the Earth as a mobile, ever-changing planet. In 1968 he encouraged colleagues to adopt the theory, writing in "Static or Mobile Earth: The Current Scientific Revolution," "Imagine a school of scientists who studied whirlpools and also held the fixed doctrine that the water in whirlpools did not move. It would not matter how detailed their studies of the topography of the surface of the water, of its chemistry or of its physical properties; they would never understand whirlpools until they changed their doctrine and admitted that the water was moving."

Newfoundlander Harold S. (Hank) Williams studied under Tuzo Wilson at the University of Toronto and eventually joined the faculty at Memorial University. Throughout the 1960s and 1970s, as plate tectonic theory gained acceptance, the geological evidence that Williams, his colleagues, and students assembled from Newfoundland sent a buzz through the geological community. Their work helped prove that plate tectonics had operated throughout much of geologic time, as Wilson had proposed. At the end of each field season, geologists working in other parts of the globe were keen to know what had been discovered in Newfoundland. The island was truly on the leading edge of a scientific revolution. One of the major accomplishments from this period of intensive research was Williams's map "Tectonic Lithofacies of the Appalachian Orogen," published in 1978 (see facing page).

## Two Supercontinents and an Ocean

Ongoing work in Newfoundland continues to contribute to geologists' understanding of plate tectonic processes and of how the Appalachian mountains evolved. Two supercontinents serve as bookends to the mountain belt's history: It begins with the destruction of the supercontinent Rodinia and ends with the creation of a new supercontinent, Pangaea. Between the two is the story of an ocean called Iapetus.

A series of continental collisions brought all the Earth's landmasses together to form the supercontinent Rodinia about 1,100 million years ago. Rodinia lay along the equator and was surrounded by a single worldwide ocean. Down the middle of the supercontinent ran a mountain belt, a part of which is now preserved as the Grenville province of the Canadian shield.

About 750 million years ago, Rodinia began to split apart. The fragmentation of Rodinia continued for more than 150 million years. As part of that process, molten rock filled cracks in the rifting continental crust, resulting in clusters of dykes that can still be seen from Newfoundland to Virginia. A rift valley formed, and as plate motion continued to pull Rodinia apart, the valley deepened. Eventually the continental crust underlying the valley collapsed, and new basaltic ocean crust began to form in the widening gap. The Iapetus ocean basin was born.

On either side of the new ocean lay diverging halves of Rodinia, which geologists call Laurentia and Gondwana. As the ocean widened, the continental margins of Laurentia and Gondwana cooled, subsided, and stabilized as they moved farther from the hot, tectonically active mid-ocean ridge. This period in the history of the Iapetus ocean is called the rift-drift transition.

For about 60 million years, the Iapetus ocean continued to open. Thick layers of limestone accumulated in a tropical climate along the stable margin of Laurentia. Remnants of this feature, called a carbonate platform, can be found in western Newfoundland and also in the southeastern US.

The supercontinent Rodinia.

The term "Iapetus ocean" suggests it was a single, simple ocean basin. But the real story is more complicated. Iapetus was probably similar to areas of the west and southwest Pacific ocean today. There, the main Pacific ocean plate has interacted with the Asian and Australian plates, creating volcanic arcs and small ocean basins of several ages, as well as numerous subduction zones and spreading centres.

Similarly, "Gondwana" was not a single coherent continental mass, but rather a complex region. At the present day, "Asia" is used to refer to an entire region that includes large island chains such as Japan and Indonesia. "Africa" includes Madagascar, although geologically many consider it a separate microcontinent. In a similar way, the name "Gondwana" is used in this field guide to mean a complex region including islands and microcontinents as well as larger regions of old, stable continental crust. The margin of Laurentia was also complex. Both regions evolved continuously as Iapetus opened and closed.

About 470 million years ago, slabs of Iapetus sediments, ocean crust, and even mantle were pushed onto Laurentia and Gondwana, forming allochthons now preserved in Newfoundland, Quebec, and New England. Around this time, Iapetus began to close more rapidly as new subduction zones formed on both sides of the ocean. Closure was hastened by subduction of the Iapetus mid-ocean ridge around this time.

The supercontinent Pangaea.

Subduction zones in the Iapetus ocean crust created new bits of land – volcanic arcs in the ocean basin. Fragments and blocks of continental crust rifted from Gondwana and drifted away, opening up new seaways and forming separate microcontinents. Closure of the main ocean basin was followed by closure of smaller seaways and basins, involving the volcanic arcs and microcontinents in complicated plate tectonic interactions. One after another, the smaller pieces were added to Laurentia.

As part of this complex process, about 425 million years ago, the Dunnage and Gander zones were subjected to great wrenching movements. The irregular shape of the colliding terrains created a complicated pattern of rifting, thrusting, and tearing of the crust. Intense metamorphism and even melting occurred near the base of the colliding continents, and granitic magma rose up through the weakened region.

By 300 million years ago, the Earth's landmasses were once again assembled into a supercontinent. Pangaea had formed. Beginning about 175 million years ago, the modern Atlantic ocean began to open. The new ocean formed approximately, but not exactly, where the Iapetus ocean had been. For that reason, parts of Laurentia and Gondwana are found on both sides of the Atlantic ocean today.

# Resources

## Museums & Interpretive Centres

Newfoundland offers a variety of museums and interpretive centres with geology-themed exhibits, including:

Baie Verte Peninsula Miner's Museum, Route 410, Baie Verte.

Bell Island Community Museum and No. 2 Mine Tour, Compressor Hill, Wabana.

Bird Cove Interpretation Centre, Brig Bay Road, Bird Cove.

Buchans Miner's Museum, Court Road, Buchans.

Dorset Soapstone Quarry National Historic Site and Museum, Route 410, Fleur de Lys.

Dover Fault Interpretation Centre, Wellington Road, Dover.

Edge of Avalon Interpretive Centre (includes Mistaken Point), Route 10, Portugal Cove South.

Fortune Head Interpretation Centre, Bunker's Hill Road, Fortune.

Gros Morne Discovery Centre, Route 431, Woody Point.

Gros Morne Visitor Centre, Route 430, Rocky Harbour.

Johnson Geo Centre, Signal Hill Road, St. John's.

Manuels River Linear Park Chalet, Route 60, Conception Bay South. (Construction of an interpretive centre is in progress.)

Pilley's Island Heritage Museum, Harbour View Road, Pilley's Island.

St. Lawrence Miner's Memorial Museum, Route 220, St. Lawrence.

## Books

Allaby, Michael. *A Dictionary of Earth Sciences*. Oxford University Press, 2008.

Attenborough, David. *David Attenborough's First Life*. HarperCollins, 2011.

Brasier, Martin. *Darwin's Lost World: The Hidden History of Animal Life*. Oxford University Press, 2009.

Colman-Sadd, Stephen P. and Susan A. Scott. *Newfoundland and Labrador Traveller's Guide to the Geology: Guidebook to Stops of Interest*. Geological Association of Canada, 1994. (The set includes a geological map and paperback guidebook.)

Eyles, Nick and Andrew Miall. *Canada Rocks: The Geologic Journey*. Fitzhenry & Whiteside, 2007. (Chapter 6, "Building Eastern Canada," describes the evolution of Newfoundland as part of the Appalachian mountain belt.)

Fortey, Richard. *Earth: An Intimate History*. UK General Books, 2005. (Chapter 6, "Ancient Ranges," is mostly about Newfoundland and related rocks in northern Europe.)

Jukes, Joseph Beete. *Excursions In and About Newfoundland, During the Years 1839 and 1840*. John Murray, 1842. (Available online at books.google.ca.)

Martin, Wendy. *Once Upon a Mine: Story of Pre-Confederation Mines on the Island of Newfoundland*. Special Volume no. 26. Quebec: Canadian Institute of Mining and Metallurgy, 1983.

Murray, Alexander and James P. Howley. *Geological Survey of Newfoundland*. Edward Stanford, 1881. (Reports for 1864-1879; available online at books.google.ca.)

## Websites

### Geology

Geological Association of Canada, Newfoundland and Labrador Section, www.gac-nl.ca (click Publications for access to online GAC-NL field guides).

Newfoundland and Labrador Geological Survey, www.nr.gov.nl.ca/nr/mines (click Geoscience).

Newfoundland and Labrador Palaeontological Resource Regulations, www.assembly.nl.ca/Legislation/sr/Regulations/rc110067.htm.

### Maps

MapsNL (topographic maps of Newfoundland and Labrador), www.mapsnl.ca.

Natural Resources Canada GeoGratis portal (free downloadable geospatial data), geogratis.cgdi.gc.ca.

Newfoundland and Labrador GeoScience Atlas OnLine (interactive geologic maps and data), gis.geosurv.gov.nl.ca/resourceatlas.

### Touring and Hiking

Canadian Hydrographic Service (tide tables), www.tides.gc.ca.

East Coast Trail Association (hiking tips), www.eastcoasttrail.ca/trail/hiking_tips.php.

Leave No Trace, www.leavenotrace.ca.

Newfoundland and Labrador Parks and Natural Areas, www.env.gov.nl.ca/env/parks.

Newfoundland and Labrador Tourism, www.newfoundlandlabrador.com.

TrailPeak (trail maps and information), trailpeak.com.

# Trip Planner

These two pages provide an easy way to plan your geological excursions. The trip planner lists the book's 48 sites – as well as related museums and interpretive centres – in geographic order. The list begins in Port aux Basques and ends in St. John's, following the Trans-Canada Highway (Route 1) from west to east. Sites can all be accessed from the province's Scenic Touring Routes as indicated in the Trip Planner. The Scenic Touring Routes are listed in order of their departure from the Trans-Canada Highway from west to east. Within each scenic route, the sites are listed in a logical travel order, but of course many variations are possible.

For more information about the province's Scenic Touring Routes, visit Newfoundland and Labrador's tourism website at www.newfoundlandlabrador.com or consult your copy of the current Newfoundland and Labrador *Traveller's Guide*.

| No. | Site | NL Scenic Touring Route | Hwy | Hike | Fac. | Rocks |
|-----|------|------------------------|-----|------|------|-------|
| 29 | Port aux Basques | Rose Blanche Lighthouse Scenic Drive | 1 | 3.5 | P | m |
| 28 | Margaree | Rose Blanche Lighthouse Scenic Drive | 470 | 0.2 | | m |
| 30 | Rose Blanche | Rose Blanche Lighthouse Scenic Drive | 470 | 1 | | i |
| 12 | Blanche Brook | French Ancestors Route | 460 | 1.3 | | s, f |
| 6 | Piccadilly Head | French Ancestors Route | 463 | 0.5 | | s |
| 5 | Mainland | French Ancestors Route | 463 | 2 | | s |
| 10 | Mine Brook | Captain Cook's Trail | 450 | 0 | | i |
| 9 | Bottle Cove | Captain Cook's Trail | 450 | 1.25 | | i, m |
| 11 | Gros Morne Discovery Ctr. | Viking Trail | 431 | 0 | M | |
| 11 | Tablelands | Viking Trail | 431 | 4 | P | i |
| 7 | Gros Morne Visitor Ctr. | Viking Trail | 430 | 0 | M | |
| 7 | Rocky Harbour | Viking Trail | 430 | 1 | P | s |
| 3 | Green Point | Viking Trail | 430 | 0.8 | P | s, f |
| 1 | Western Brook Pond | Viking Trail | 430 | 5.4 | P | i, m |
| 4 | Table Point | Viking Trail | 430 | 2.5 | P | s, f |
| 2 | Bird Cove Interpretation Ctr. | Viking Trail | 430 | 0 | M | |
| 2 | Flowers Cove | Viking Trail | 430 | 1.4 | | s, f |
| 8 | L'Anse aux Meadows | Viking Trail | 430 | 0.25 | | s |
| 18 | Baie Verte Pen. Miner's Museum | Dorset Trail | 410 | 0 | M | |
| 19 | Coachman's Cove | Dorset Trail | 410 | 0.1 | | m |
| 18 | Dorset Soapstone Museum | Dorset Trail | 410 | 0 | M | |
| 18 | Fleur de Lys | Dorset Trail | 410 | 0.25 | P | m |
| 22 | Goodyear's Cove | Green Bay and the Beothuk Trail | 1 | 0.75 | P | i |
| 16 | Pilley's I. Heritage Museum | Green Bay and the Beothuk Trail | 380 | 0 | M | |
| 16 | Buchans Miner's Museum | Exploits Valley | 370 | 0 | M | |
| 16 | Millertown | Exploits Valley | 370 | 0.5 | | i |
| 24 | Bishop's Falls | Exploits Valley | 1 | 0.1 | P | s |
| 25 | Leading Tickles | Exploits Valley | 350 | 0 | | i |
| 17 | Trout Hole Falls | Coast of Bays | 361 | 0.25 | P | s |
| 20 | Summerford | Kittiwake Coast: Road to the Isles | 340 | 1.5 | | s, m |
| 15 | Moreton's Harbour | Kittiwake Coast: Road to the Isles | 345 | 1 | | i |

For your convenience, the Trip Planner lists the following information:

**Site name and number** (see Table of Contents and each site heading)

**Scenic Route** (the provincial Scenic Touring Route on which the site is located)

**Nearest Highway** (the highway pictured in the road map for the site)

**Hiking Distance** (the round-trip distance between the parking location and the outcrop, in kilometres)

**Facilities** (P, located in a national, provincial, or municipal park or reserve; M, the listed site is a museum or interpretive centre)

**Rock Types** (i, igneous; s, sedimentary; f, fossils; m, metamorphic)

| No. | Site | NL Scenic Touring Route | Hwy | Hike | Fac. | Rocks |
|-----|------|-------------------------|-----|------|------|-------|
| 21 | Pike's Arm | Kittiwake Coast: Road to the Isles | 346 | 0.5 | | s |
| 14 | Little Harbour (Twillingate) | Kittiwake Coast: Road to the Isles | 340 | 0.1 | | i |
| 13 | Sleepy Cove | Kittiwake Coast: Road to the Isles | 340 | 0.25 | | i |
| 23 | Tilting, Fogo | Kittiwake Coast: Islands Experience | 334 | 1.75 | | i |
| 26 | Gander | Kittiwake Coast: Road to the Shore | 1 | 0.5 | | s |
| 27 | Little Harbour (Gander Lake) | Kittiwake Coast: Road to the Shore | 1 | 0.2 | P | m |
| 34 | Lumsden | Kittiwake Coast: Road to the Shore | 330 | 1.5 | | i |
| 31 | Windmill Bight | Kittiwake Coast: Road to the Shore | 330 | 1 | P | m |
| 32 | Greenspond | Kittiwake Coast: Road to the Shore | 320 | 2 | | i |
| 33 | Dover Interpretation Ctr. | Kittiwake Coast: Road to the Shore | 320 | 0 | M | |
| 33 | Dover | Kittiwake Coast: Road to the Shore | 320 | 0.8 | | m |
| 40 | Traytown | Kittiwake Coast: Road to the Beaches | 310 | 0 | | i |
| 45 | Keels | Discovery Trail | 235 | 0.3 | | s |
| 35 | Burin | Heritage Run | 221 | 0.1 | | i |
| 48 | St. Lawrence | Heritage Run | 220 | 2 | | i |
| 48 | Miner's Memorial Museum | Heritage Run | 220 | 0 | M | |
| 43 | Fortune Head | Heritage Run | 220 | 0.8 | P | s, f |
| 43 | Fortune Head Interpretation Ctr. | Heritage Run | 220 | 0 | M | |
| 47 | Point Lance | Cape Shore | 100 | 0.4 | | i |
| 42 | Cataracts | Cape Shore | 91 | 0.25 | P | s |
| 38 | Mad Rock | Baccalieu Trail | 70 | 2 | | s |
| 37 | St. Mary's | Irish Loop | 90 | 2 | | s |
| 39 | Edge of Avalon Interpretive Ctr. | Irish Loop | 10 | 0 | M | |
| 39 | Mistaken Point | Irish Loop | 10 | 6 | P | s, f |
| 44 | Bacon Cove | Admiral's Coast | 60 | 0.4 | | s, m |
| 36 | Manuels River Interpretive Ctr. | Admiral's Coast | 60 | 0 | M | |
| 36 | Manuels River | Admiral's Coast | 60 | 2 | P | i, s |
| 46 | Bell Island | Killick Coast Trail | 40 | 0.2 | | s, f |
| 46 | Bell Island No. 2 Mine Tour | Killick Coast Trail | 40 | 0.25 | M | |
| 41 | Johnson Geo Centre | St. John's & Environs | 30 | 0 | M | |
| 41 | Signal Hill | St. John's & Environs | 30 | 4.5 | P | s |

# HUMBER
## *Zone at a Glance*

### Boundaries

West: Logan's Line
East: Baie Verte-Brompton Line

### Origin

Continental margin of Laurentia

### Characteristic Features

Ancient gneisses
Carbonate shelf
Melanges
Displaced ocean crust

## In the Humber zone, you can …

| | | |
|---|---|---|
| **1** | **Western Brook Pond** | Take a boat tour through ancient gneisses. |
| **2** | **Flower's Cove** | Share the shoreline with primitive life forms of a tropical ocean. |
| **3** | **Green Point** | Stand on an international boundary of the geologic time scale. |
| **4** | **Table Point** | Visit creatures that lived in an ancient ocean. |
| **5** | **Mainland** | Experience the collapse and burial of a continental margin. |
| **6** | **Piccadilly Head** | Witness the arrival of a colliding plate. |
| **7** | **Rocky Harbour** | Explore the chaos within a tectonic melange. |
| **8** | **L'Anse aux Meadows** | See how the ocean floor travelled onto the land. |
| **9** | **Bottle Cove** | View the remains of an ocean island. |
| **10** | **Mine Brook** | Study a volcanic pillow from the ocean floor. |
| **11** | **Tablelands** | Walk on the Earth's mantle, once 8 kilometres beneath the ocean floor. |
| **12** | **Blanche Brook** | Discover a fossil forest of giant trees. |

The Humber zone contains Newfoundland's oldest rocks. In fact, 1,000 million years ago the granitic rock of Newfoundland's Long Range (site 1-1) was the only part of Newfoundland in existence. That little bit of Newfoundland lay beneath an even older mountain range that has since been worn away, a mountain range that cut across the supercontinent geologists call Rodinia.

Around 610 million years ago the rifting of Rodinia (site 1-2) caused cracks to appear and fill with hot lava from beneath the continent. Next, rift valleys formed and widened, filling with sediments eroded from the steep valley walls.

As tectonic plate movements continued, the rift valley became deeper. Sea water flooded in and eventually a new ocean opened – the Iapetus ocean. The Humber

Western Brook Pond (see site 1).

zone now lay on the edge of a new continent, Laurentia, along Iapetus's shore. Between 540 and 470 million years ago, a carbonate platform accumulated on the quiet, shallow continental shelf of Laurentia (site 2). At the same time, on the continental slope, sedimentary layers unique to that deep water environment formed also (site 3). For about 70 million years, the shores of Laurentia were tectonically peaceful as new life forms evolved in the ocean.

Green Point (see site 3).

Winterhouse Brook (see site 11).

Beginning about 470 million years ago, plate tectonic movements changed direction, and the Iapetus ocean began to close. A slab of the ocean floor wrenched upward and moved toward Laurentia. Its approach bent the continental margin, creating first a massive crustal bulge and then a deep foreland basin. In the Humber zone, this caused an unconformity (site 4) followed by collapse of the carbonate platform and a flood of new sediment eroded from the approaching oceanic slab (site 5).

By 460 million years ago, closure of the Iapetus ocean had left two huge stacks of oceanic rocks on top of the Humber zone in the form of the Humber Arm and Hare Bay allochthons. They consist of a series of thrust slices, each riding on a layer of melange. The lower slices and melanges contain ocean sediments (sites 6, 7, and 8), while the upper slices and melanges contain ocean crust and mantle (sites 9, 10, and 11).

Throughout the rest of the Appalachian orogeny, deformation affected the rocks of the Humber zone. Very few new rocks were formed until the Carboniferous period, when a long narrow basin formed along the eastern edge of the Humber zone. Once the site of primitive, tall forests, lakes, and streams, the basin later filled with sediment that still covers parts of the zone (site 12).

Blanche Brook (see site 12).

Newfoundland's Long Range mountains as seen from a picnic area near the tour-boat terminal at Western Brook Pond, Gros Morne National Park.

# Beneath Older Mountains

## Proterozoic Gneisses at Western Brook Pond

Few roads cross it. No towns populate its barren heights. Grand and remote, the Long Range of the Great Northern peninsula is a world apart – and its origins are as distant in time as its peaks are distant from view.

The Long Range currently forms the northeastern end of North America's Appalachian mountains. But the granitic gneisses of the Long Range once lay deep beneath an older mountain chain, at the heart of an ancient supercontinent geologists call Rodinia (see The Appalachian Mountains, page 17).

The boat tour at Western Brook Pond in Gros Morne National Park provides a unique journey among granitic gneisses that formed 1,500 million years ago. They lay deep in the Earth 1,000 million years ago as continents collided to form Rodinia.

You'll also see evidence of Rodinia's demise – rocks filling cracks that formed as the supercontinent began to split apart about 600 million years ago.

# Getting There

### Driving Directions

Follow Route 430 to Gros Morne National Park. About 11 kilometres south of St. Paul's, watch for the parking area on the east side of the road. There are road signs for Western Brook Pond on approach from either direction.

### Where to Park

Parking Location: N49.78719, W57.87490

Enter the large parking lot directly from Route 430.

### Walking Directions

A well-marked and well-maintained path of alternating boardwalk and gravel sections leads from the parking area to the tour-boat terminal (N49.78124, W57.84132) on Western Brook Pond. A few additional paths branch off for other hikes and related camping areas; follow the signs for the boat tour.

### Notes

The hike offers many great views and informative interpretive panels along the way, so plan plenty of time for your hike. You may want to make a reservation, since tour-boat capacity is limited. A pair of binoculars can be helpful for viewing details during the boat tour. You must have a valid Parks Canada pass to hike the trail, and there is a fee for the tour itself. For more information, visit www.pc.gc.ca/grosmorne.

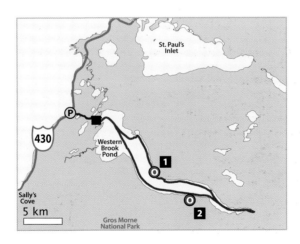

---

**1:50,000 Map**

Gros Morne 012H12

St. Paul's Inlet 012H13

**Provincial Scenic Route**

Viking Trail

# On the Outcrop (1)

Granitic gneisses dominate the north shore of Western Brook Pond.

**Outcrop Location: N49.74376, W57.77707**

The terminal for the Western Brook Pond boat tour is located in the same low-lying landscape you hike across to get there. Where the pond narrows, there is a sharp contrast in topography as the hills abruptly rise from the coastal plain along the Long Range fault.

The hike and open reaches of the boat tour cross layers of Paleozoic limestone and shale, as explained on the park's interpretive panels. When the boat enters the steep-walled fjord, though, you come face to face with Newfoundland's oldest rocks, part of a terrain known to geologists as the Long Range inlier. The outcrop location noted above is a point on the tour route near the first large cliffs of granitic gneiss.

On the tour, you'll pass many bare rock surfaces. Some are stained by water seepage or weathering, making it difficult to see the gneiss features, but on the freshest outcrops you can see them – wispy, streaky bands of grey across a lighter pinkish background. Occasionally clumps or knots of darker grey or pink are visible. Several recent landslides have exposed the freshest surfaces.

# FYI

- The gneisses at Western Brook Pond formed about 1,500 million year ago and were metamorphosed about 1,000 million years ago.

- Rocks of the Long Range lay deep in the Earth's crust until some time after 430 million years ago, when they were pushed up along the Long Range fault.

- The Long Range inlier is part of the Grenville province of the Canadian shield. Gneisses with a similar history can be found in northern Ireland and Scotland.

## Related Outcrops

For a closer look at the gneisses, a fresh outcrop can be examined adjacent to a pull-off about 8.5 kilometres north of Wiltondale on Route 430 in Gros Morne National Park (N49.46212, W57.64622).

Erin mountain in Barachois Pond Provincial Park is part of an anorthosite of similar age, about 1,200 million years old. The park's Erin Mountain Trail begins at the bridge across the narrows (N48.48128, W58.26035).

Precambrian gneiss along Route 430 in Gros Morne National Park.

In Newfoundland, rocks 1,000 million years or more in age (shaded red in the map at left) occur only in the Humber zone. They underlie most of the Great Northern peninsula and form smaller isolated blocks further south.

## Exploring Further

Atlas of Canada website, atlas.nrcan.gc.ca. Search "geological provinces" to see maps of rock ages in Canada.

| 500 | | 400 | | 300 | | 200 | | 100 | | 0 |
|---|---|---|---|---|---|---|---|---|---|---|
| € | O | S | D | C | P | ₸ | J | K | Cz | |

# On the Outcrop (2)

Dark, near-vertical diabase dykes cutting granitic gneisses along the southern shore of Western Brook Pond: (a) a dyke visible deep within a recess in the cliff face at the outcrop location; (b) about 2 kilometres farther west, a dyke cutting a bare surface exposed by a recent landslide.

**Outcrop Location: N49.72279, W57.73644**

Long after the granitic gneisses of the Long Range had formed and cooled, plate tectonic forces began to pull the ancient continent of Rodinia apart. Molten rock from the Earth's hot interior flowed into the resulting cracks, forming diabase dykes that now cut across the landscape.

Diabase is rich in the dark mineral pyroxene, so the rock often appears black or dark grey in the field. The dykes are approximately vertical, and their parallel sides form sharp contacts with the lighter pinkish grey gneisses.

The boat tour normally pauses to point out one or more of these dark gashes in the cliffs of Western Brook Pond. The two well-exposed examples pictured above are visible on the south shore.

1000      900      800      700      600      500

$Z_1$      $Z_2$      $Z_3$      $\in$

# FYI

- The diabase dykes at Western Brook Pond are part of a large dyke swarm. The swarm covers much of the Great Northern peninsula and is one of several in eastern Canada that formed during the late Proterozoic era.

- Rifting of Rodinia was followed by a rift-drift transition when great volumes of sediment accumulated, first in rift valleys and then in the new, narrow ocean basin. Such sediments form important landmarks in the Humber zone, including North Summit and South Summit near Barr'd Harbour and Gros Morne mountain near Rocky Harbour.

The sedimentary layers visible on the North Summit near Barr'd Harbour were deposited in a young, narrow Iapetus ocean soon after it opened.

## Related Outcrops

On the shoreline along Green Gardens Trail in Gros Morne National Park is pillow lava also formed during the rifting of Rodinia. They are slightly younger than the dykes (550 million years old) but formed as part of the same process that opened the Iapetus ocean.

| 500 | | | 400 | | 300 | | 200 | | 100 | | 0 |
|---|---|---|---|---|---|---|---|---|---|---|---|
| Є | O | S | D | C | P | Tr | J | K | | Cz | |

Thrombolites emerge as the tide falls in Flower's Cove.

# Tropical Paradise
## Thrombolite Mounds at Flower's Cove

When the Iapetus ocean opened and began to widen, western Newfoundland was on the edge of a new continent, Laurentia. Its stable continental margin was free from crustal upheaval or volcanic activity. Plate tectonic movements had positioned the region near the equator – expanses of warm ocean supported many new life forms.

Thrombolites are limestone mounds formed by colonies of ocean-dwelling microbes. They were common at this point in Earth history, first appearing early in the Cambrian period. By the end of the Cambrian period, they had reached the peak of their success.

Throughout the Ordovician period their abundance declined, and they all but disappeared as other life forms took over the "mound business," building structures better able to resist burrowing and predation by an evolving army of new marine life (for example, see site 4).

At Flower's Cove many of the fossil thrombolites are completely exposed at low tide. But when they were alive long ago, they were probably in slightly deeper water, so that at low tide the water just covered their broad, flat tops.

# Getting There

### Driving Directions

Follow Route 430 to Flower's Cove. Watch for road signs to the turn-off (N51.28733, W56.73973) for Thrombolites Walking Trail, which is on the south side of town.

### Where to Park

Parking Location: N51.28798, W56.74092

Park in the lot designated for visitors to the Thrombolites Walking Trail.

### Walking Directions

Follow the boardwalk and Marjorie Bridge to a gravel hiking trail along the shoreline. Walk along the trail to a viewing/rest area.

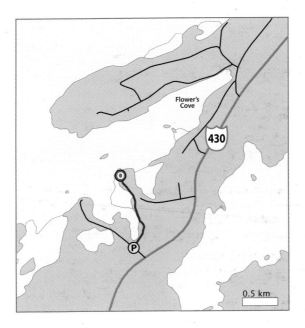

---

**1:50,000 Map**

Flower's Cove 012P07

**Provincial Scenic Route**

Viking Trail

# On the Outcrop

Cambrian thrombolite mounds merged into a large group.

**Outcrop Location: N51.29350, W56.74253**

Emerging like a tray of gigantic macaroons from the shallow water of the cove, the thrombolites of Flower's Cove form mounds 40 to 200 centimetres or more in diameter.

Many kinds of marine life create mound-like structures (for example, the corals of our modern oceans). As life evolved, organisms developed the ability to form exoskeletons – hard, protective shells – with distinctive shapes for each individual species. The Cambrian thrombolites represent a more primitive strategy and were built up of small clumps of calcified microbe colonies.

Superficially, thrombolites resemble stromatolites (see site 45) in that both form mounds, and both were common in the Cambrian period. But stromatolites are built up of distinct layers, often visible as fine concentric lines of sediment within the mound. As you'll see, there is no such layering within the mounds at Flower's Cove.

Like modern coral reefs, thrombolites provided a habitat for many other organisms living within and among the mounds. The most obvious evidence for this is the numerous burrows that puncture the mounds.

| 1000 | 900 | 800 | 700 | 600 | 500 |
|------|-----|-----|-----|-----|-----|
| | $Z_1$ | | $Z_2$ | $Z_3$ | $\in$ |

# FYI

- Thrombolites probably lived in shallow water, for example, in lagoons or tidal flats.

- Modern thrombolite-like mounds do exist. However, they are rare, and paleontologists disagree about whether they are equivalent to Cambrian varieties.

Spaces between groups of thrombolite mounds may represent channels where water flowed as the tide rose and fell.

## Related Outcrops

A warm, shallow marine environment lasted for millions of years in the Iapetus ocean along the stable margin of Laurentia, so Cambrian and Ordovician limestones (shaded red in the map at left) are widely distributed in the Humber zone.

From the Bird Cove Interpretation Centre (N51.05275, W56.93075) you can access several kilometres of trails on the nearby Dog peninsula and explore the rocks from this time period.

## Exploring Further

Bird Cove Interpretation Center website, www.bigdroke.ca.

Knight, I. and W.D. Boyce. *A Short History of the Geology of the Bird Cove Area: An Educational Resource and Field Guide.* Geological Survey of Newfoundland and Labrador, 2003. (Available online.)

Looking north to Green Point from the adjacent campground in Gros Morne National Park.

# Time Line

## Cambrian-Ordovician Boundary at Green Point

It took twenty-six years, two working groups, fifteen ballots, and three rounds of approval: In January 2000, the International Union of Geological Sciences designated Green Point in Gros Morne National Park as a Global Stratotype Section and Point (GSSP). That's an official international reference point for the geologic time scale.

The site certainly looks like reference material – walking along the beach toward the cliffs of Green Point is like entering the pages of a book that's three storeys tall.

One of the most intensively scrutinized outcrops in the world, Green Point provides an unbroken record of millions of years of sediment accumulation in the Iapetus ocean. The rocks formed in deep water far from shore, near the base of the continental slope.

Researchers and students at Memorial University and elsewhere studied more than 10,000 fossils from Green Point, using tiny conodonts and delicate graptolites to locate the boundary between the Cambrian and Ordovician periods. They've narrowed it down to a single layer, Bed 23, that contains the first appearance of a conodont, *Iapetognathus fluctivagus*.

# Getting There

### Driving Directions

Follow Route 430 to Gros Morne National Park. About 9.5 kilometres north of Rocky Harbour, you'll see two road signs for Green Point – one for a campground and, just north of the campground, another (N49.68264, W57.95626) for the entrance to the GSSP. Turn west onto the gravel road and follow it to the parking area.

### Where to Park

Parking Location: N49.68256, W57.96110

The parking lot for visitors to the GSSP is on the high ground above the beach.

### Walking Directions

From the parking lot, walk along the gravel road down the hill to the shore. Follow the path north along the shore; continue north along the boulder beach and around a prominent rock layer to the shore immediately west of the headland.

### Notes

There is no fee specific to the GSSP, but you must have a valid Parks Canada pass to visit the site (see www.pc.gc.ca/grosmorne for details). This site protects an international research and reference sequence of rock layers. Please respect its importance.

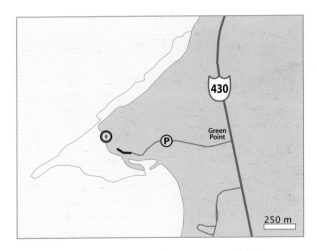

---

**1:50,000 Map**

Gros Morne 012H12

**Provincial Scenic Route**

Viking Trail

# On the Outcrop

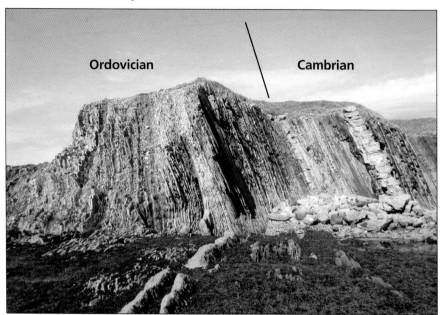

Ordovician      Cambrian

This west-facing cliff face at Green Point, Gros Morne National Park, contains the Cambrian-Ordovician Global Stratotype Section and Point.

## Outcrop Location:  N49.68307, W57.96544

Best viewed at low tide and with the sun in the west, the GSSP at Green Point contains steeply tilted beds of shale and limestone. Cambrian rocks outcrop in the southern half of the cliff face and Ordovician rocks in the northern half. Bed 23, the boundary layer, occupies a small recess in the headland (see photo annotation, above).

Shale layers (photo a) are black, grey, or green. Black layers contain more organic material, while green layers formed when oxygen levels were low on the sea floor. Such variations might be due to cycles of climatic change.

In among the shale layers are two very different types of limestone. You'll walk around a single wide bed of limestone conglomerate (photo b) on your way to view the time boundary. A sort of underwater landslide brought the conglomerate fragments down the sloping sea floor to form this layer. Thin, light-coloured layers (photo c) are ribbon limestones. The buckling in them might have formed as the sediments were being compressed into rock or could have resulted from the tectonic forces that tilted the rocks millions of years later.

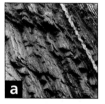

a

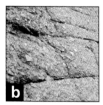

b

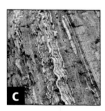

c

| 1000 | 900 | 800 | 700 | 600 | 500 |
|------|-----|-----|-----|-----|-----|
| | $Z_1$ | | $Z_2$ | $Z_3$ | $\in$ |

# FYI

- The International Union of Geological Sciences considered sites from Kazakhstan, China, Australia, Scandinavia, Britain, and the US before selecting Green Point as the best site for the GSSP.

- Life forms with shells, such as trilobites and brachiopods, existed during this time period, but they preferred a shallow-water habitat. Since Green Point was a deep-water environment, shelly fossils are rare here.

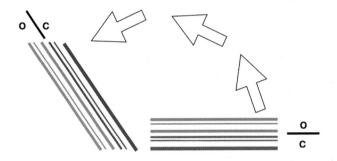

Normally, younger rock layers lie horizontally above older ones. However, the rock layers at Green Point were overturned into their present tilted position.

## Related Outcrops

The rocks at Green Point formed far offshore to the east and were pushed onto the Humber zone as part of the Humber Arm allochthon. The area occupied by these and related sediments and their underlying melange is shaded red in the map at right.

The Precambrian-Cambrian GSSP is also located in Newfoundland, at Fortune Head on the Burin peninsula (see site 43).

## Exploring Further

International Commission on Stratigraphy website, www.stratigraphy.org.

Inclined beds of Ordovician limestone along the shore at Table Point Ecological Reserve.

# Radiating Life

## Ordovician Fossils at Table Point Ecological Reserve

Dry, angular chunks of broken limestone crunch underfoot as you walk across the rocky desolation of Table Point. Not much lives out on the headland – you'll see a bit of tuckamore in a gully, perhaps a solitary insect or a few seabirds.

Yet for millions of years Table Point was the site of a thriving ecosystem along the tropical shores of the Iapetus ocean. The Ordovician period saw a sudden increase in the diversity of life forms – an evolutionary radiation. During the middle of the Ordovician period this area hosted abundant species of gastropods, trilobites, and brachiopods as well as sponges, crinoids, cephalopods, and others.

Table Point and its fossils have drawn the attention of researchers since Elkanah Billings first described dozens of species from this site in his work *Paleozoic Fossils*, published in 1865. To protect such a unique resource, the site was designated as a provincial ecological reserve in 1990.

The rocks at Table Point are interesting in another way, too. As they formed, new tectonic forces warped the once-stable continental margin of Laurentia. The Iapetus ocean was closing and change was on the way (see site 5).

# Getting There

## Driving Directions

The ecological reserve lies immediately west of Route 430, beginning about 3 kilometres north of Bellburns and extending a little more than 2 kilometres farther north.

There are no road signs indicating the location of the reserve, and vehicular access is prohibited.

## Where to Park

No parking area is provided specifically for the reserve, but you can pull off the road at two sites nearby:

N50.34482, W57.52931

About 1 kilometre south of the reserve, in a picnic area west of Route 430.

N50.37493, W57.52465

About 0.5 kilometres north of the reserve, west of Route 430 near the intersection with a gravel road.

## Walking Directions

A trail loops into the reserve from N50.36562, W57.52911 and from N50.35380, W57.52822. Follow it onto the barren headland, then walk across the rocky landscape toward the water.

## Notes

Table Point Ecological Reserve and its fossils are protected by law. It is illegal to disturb or remove any fossils, rock materials, or other natural features.

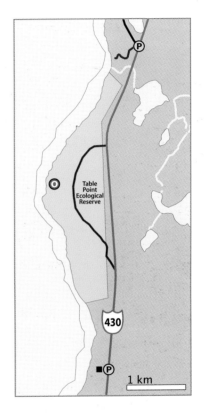

---

**1:50,000 Map**

Belburns 012I05-06

**Provincial Scenic Route**

Viking Trail

# On the Outcrop

The tilted upper surface of a limestone layer is exposed in a gully at Table Point Ecological Reserve. Such gullies expose a wide variety of fossils, for example, (a) a mineralized gastropod; (b) a gastropod filled with other shell debris; (c) a brachiopod.

## Outcrop Location: N50.36248, W57.53656

Because of the way the limestones at Table Point are tilted, as you walk from north to south the rock layers become younger. The environment in which they formed changes, too, from shallow tidal flat conditions in the north, through lagoon and shoal conditions, to deep-water conditions in the southern parts of the reserve.

Many fossils can be seen in the cliff face and its rocky base along the shore at Table Point. But up on the headland, rain washes surface rubble away in a series of gullies that arc toward the shore, exposing the tilted upper surface of certain layers. Access on this high ground is less dependent on tides and weather. Fossils are easy to spot in exposures such as this one.

You don't have to be a paleontologist to enjoy a collection so varied in appearance. Some fossils are filled with sparkling pyrite (photo a), others with bright white calcite. Some are mere impressions in the rock surface, but in others the shells are preserved with such precision a bird might be tempted to peck at them for a juicy morsel.

1000    900    800    700    600    500

| $Z_1$ | | $Z_2$ | | $Z_3$ | € |

# FYI

- Modern limestones only form where the water is warm and clear. This suggests the land surface near Table Point was low-lying (not much erosion) and arid (no large rivers bringing sediment to the ocean).

- An unconformity separates the limestones of Table Point from older ones to the north. An equivalent gap in the rock record occurred in many other parts of eastern North America, from New England to Alabama, as the whole continental margin flexed up then back down (see Related Outcrops, below, and site 5).

- The same species of trilobites and gastropods found at Table Point occur in Nevada (US), as well as in parts of Siberia, Kazakhstan, and Scandinavia. Called the Toquima-Table Head Biogeographic province, these far-flung sites may mark the shores of Laurentia during the Ordovician period.

Limestones of contrasting colour appear above and below a major unconformity at Aguathuna quarry.

## Related Outcrops

The unconformity at the base of the Table Point limestones is clearly exposed at Aguathuna quarry (N48.56171, W58.77247) on the Port au Port peninsula. On the long quarry wall, the darker Table Point layers drape over the uneven, eroded surface of the lighter grey rocks below.

### Exploring Further

Billings, Elkanah. *Paleozoic Fossils*. Dawson Brothers, 1865. (Available online at books.google.ca.)

Grey shales and limestones are capped by greenish sandstone along the beach just south of Mainland, Port au Port peninsula.

# Collapse and Burial
## Impending Tectonic Collision at Mainland

It's 470 million years ago, the middle of the Ordovician period. For 40 million years limestones have formed in the clear water of the stable Laurentian continental shelf (see sites 2-4).

Say goodbye to stability – earthquakes now rattle the region. The once orderly layers of the Cambrian-Ordovician platform are fractured, fragmented, dismantled. Their broken remains are smothered with sediment – piles of sand pouring in from somewhere offshore … it's unstoppable … and it's green …

On a cobble beach south of Mainland, the scientific details of this real-life disaster plot have been told and retold to geology students, visiting researchers, and field gatherings. Meticulous research uncovered the tectonic drama recorded by the limestone, shale, and sandstone layers on view here, all caused by the approach of the Humber Arm allochthon during closure of the Iapetus ocean.

As you walk along the beach, the story unfolds before you: Evidence for the collapse of the stable platform and its subsequent burial as sediment derived from the allochthon poured into a newly formed foreland basin.

# Getting There

### Driving Directions

On the Port au Port peninsula, from Lourdes or Cape St. George, follow Route 463 to Mainland. At the south end of the town, turn west onto Caribou Brook Road and follow the gravel road to its end point along the shore.

### Where to Park

Parking Location: N48.55714, W59.18899

Park at the end of Caribou Brook Road, just above the cobble beach.

### Walking Directions

Walk from the parking area down onto the cobble beach. Follow the beach along the cliff face for about 1 kilometre until you reach a waterfall (N48.55116, W59.19773). Walk back toward Mainland, observing the rock layers until you reach the greenish sandstone (N48.55385, W59.19311).

### Notes

Walking on the large cobbles is easier if you can find a recent ATV track that has compacted the stones. There are obvious signs of rockslides along the cliff – it's unstable.

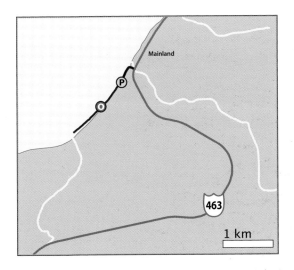

---

**1:50,000 Map**

Mainland 012B11

**Provincial Scenic Route**

French Ancestors Route

# On the Outcrop

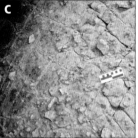

Distinctive rock types tell the story of the approaching allochthon: (a) the Mainland sandstone, thin-bedded and distinctly greenish; (b) a limestone conglomerate with abundant large boulders; (c) a limestone with typical debris flow features – isolated, randomly distributed boulders in a fine matrix.

## Outcrop Location: N48.55385, W59.19311

As you walk from the mouth of Caribou Brook northeast back toward Mainland, you'll see the rock layers in the order in which they formed – oldest near the brook, youngest near town. In the older group (Cape Cormorant formation) black shales alternate with more prominent, light grey limestones.

Most of the limestones are conglomerates containing obvious large rock fragments. The layers have features typical of debris flows, with cobbles and boulders of all sizes and shapes distributed randomly in a finer matrix.

Fossils in the conglomerate fragments have revealed an unusual pattern: the younger the rock layer, the older the fragments it contains. This occurred during unroofing of the collapsed platform, as deeper and deeper layers were exposed and eroded.

The outcrop location noted above marks the contact between the shale-limestone sequence and the younger, greenish Mainland sandstone. It is a finely layered turbidite, each layer formed as a slurry of sandy sediment swirled down a steep underwater slope into the foreland basin.

The sandstone formed from erosion of the approaching allochthon. Mineral grains and rock chips of ophiolitic material contribute to the rock's greenish colour.

1000          900          800          700          600          500

Z₁                    Z₂                    Z₃              €

# FYI

- Geologists mapping the Cape Cormorant formation along the peninsula's west coast have found that the limestone conglomerate layers become thicker and the rock fragments in them become larger toward the south. Near Big Cove, conglomerate layers more than 50 metres thick contain fragments up to 100 metres long.

- One of the most exciting discoveries made here was that the Mainland sandstone contains the mineral chromite. Chromite only occurs in rock types such as peridotite, common in the Earth's mantle but rarely found on the Earth's surface. Chromite at this site records the approach of mantle rocks as part of the Humber Arm allochthon and has helped geologists better understand plate tectonic processes.

## Related Outcrops

The Cape Cormorant formation is restricted to the Port au Port peninsula, but a narrow layer of rock similar to the Mainland sandstone occurs in the Table Point Ecological Reserve (see site 4) about 0.5 kilometres from its southern boundary. Similar sandstones occur at Lobster Cove Head (see site 7) and in the area around Hare Bay on the Great Northern peninsula. Known collectively as the Goose Tickle group (shaded red in the map at right), they were all deposited in a similar foreland basin environment.

Elsewhere on the Port au Port peninsula, for example, near Piccadilly Head, you can see other dramatic effects as the Humber Arm allochthon continued to move westward (see site 6).

500   400   300   200   100   0

€  O  S  D  C  P  Ŧ  J  K  Cz

Low cliffs containing part of the Humber Arm allochthon frame the sandy beach near Piccadilly Head, Port au Port peninsula.

# Collision Begins
## Thrust Slice near Piccadilly Head

Walking along the shore near Piccadilly Head, you can almost feel a rumble of motion. The evidence is striking: A slice of the Earth's crust travelled westward to this site.

During the middle Ordovician, the Humber Arm allochthon pushed onto Laurentia from the Iapetus ocean basin. The collision displaced rock layers along a series of low-angle thrust faults;  rock types once side by side were eventually stacked in a corresponding vertical sequence. This phenomenon is seen in many of the world's mountain ranges.

In a narrow zone along each major thrust fault, rocks were crushed and mixed to form a tectonic melange. Above and below the melange zones, rock layers in each thrust slice remained relatively undisturbed.

Several of these features can be seen along the shore near Piccadilly Head. As a bonus, on a clear day, some 40 kilometres to the northeast, displaced ocean crust in the uppermost slice is just visible atop the Lewis Hills, all part of the same collision process.

# Getting There

## Driving Directions

Follow Route 463 to the area around West Bay Centre and Piccadilly Head. About 1.25 kilometres west of the entrance to Piccadilly Head Park, pull in at the convenience store/gas bar on the north side of the road, just above the shore. A nearby rock quarry on the south side of the road will help you recognize the location.

## Where to Park

Parking Location: N48.59286, W58.91841

Park at the convenience store, being careful not to block access to the fuel pumps.

## Walking Directions

Just west of the parking lot, follow a path through the trees and along a stream to the beach. The first outcrop location is just east of the mouth of the stream. For the second outcrop location, walk farther east around the small rocky point, then continue east along the beach.

## Notes

The path from the convenience store to the beach is steep and narrow, with a small stream to cross on the shore. You can only walk around the small point at low tide. For alternative access to the beach, park in Piccadilly Head Park (open seasonally). Walk about 1 kilometre west along the beach to the outcrops.

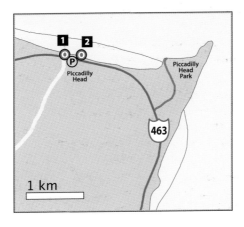

---

**1:50,000 Map**

Stephenville 012B10

**Provincial Scenic Route**

French Ancestors Route

# On the Outcrop (1)

A melange with slate matrix and numerous limestone blocks marks the base of the tectonic slice in a small cliff west of the beach at Piccadilly.

**Outcrop Location: N48.59324, W58.91858**

At this location you can see the contact between the base of a thrust slice and the continental margin it travelled over.

As you step across the stream on the shoreline, at your feet are thin layers of Table Point limestone (see FYI below and site 4), quite undisturbed. These rocks are part of the continental margin below the thrust slice.

Once you cross the stream, look east. Rising from the shore is a layer that looks like a badly made stone wall (see above). It contains blocks of limestone surrounded by distorted, strongly layered shale.

Immediately above that, somewhat obscured by brush, is a layer of dark flaky shale. In places, the shale looks like nothing more than a pile of small, loose, dark grey rock chips.

The shale, being weak and flaky, is the layer on which the thrust slice travelled. During those movements, the shale worked its way down into part of the underlying Table Point limestone and mixed with it to form a melange.

Continue east around the headland – at the next outcrop, you'll see what came along for the ride in the allochthon above the melange.

| 1000 | 900 | 800 | 700 | 600 | 500 |
|------|-----|-----|-----|-----|-----|
| $Z_1$ | | $Z_2$ | | $Z_3$ | $\in$ |

# FYI

- Table Point limestone is also found in the quarry across the road near the convenience store.

- The surface along which a thrust slice moves is called the detachment plane. Detachment planes often follow weak layers like the shale at this site or other easily pulverized rock types, and are often nearly horizontal like this one.

Table Point limestones outcrop along the mouth of the stream on the shore. The limestones lie below the melange.

## Related Outcrops

Many sites in western Newfoundland are well known for their melanges, all related to the transport of allochthons from the Iapetus ocean basin westward onto the Laurentian continent. Sites 7, 8, and 9 provide details about some of them. See site 4 for more information about the Table Point limestone.

500    400    300    200    100    0

Є O S D C P Ŧ J K Cz

# On the Outcrop (2)

Alternating layers of shale and sandstone form west-facing folds along the beach near Piccadilly Head.

**Outcrop Location: N48.59276, W58.91596**

Just east of the small point on the beach below the convenience store, for 100 metres or more along the beach, is a series of folds including the one pictured above. They are part of a lower thrust slice carrying the Humber Arm allochthon. Looking like the nose of a jet or locomotive en route, they convey a vivid sense of motion.

Forces pushing westward (left to right in the photo above) created the folds and pushed them over on their sides. The lighter-coloured sandstone, being stronger, was bent into coherent shapes, while the darker, weaker shale forms a crumbly, flaky mass between the sandstone layers.

An interesting feature of the sandstone is its graded bedding. The bottom of each layer is gritty, while the tops are of finer sand. Like the sandstones near Mainland (site 5), these are turbidites. They formed as underwater avalanches of sediment swirled down steep slopes (see FYI).

1000    900    800    700    600    500

$Z_1$    $Z_2$    $Z_3$    $\varepsilon$

# FYI

- Because the folds at this site are overturned, several of the folded layers are facing bottom-up where they meet the beach. This makes it easy to see the beautiful sole marks there (photos a and b).

- Turbidites are formed by turbidity currents, fast-moving underwater avalanches of water and sediment. As a turbidity current races down a slope, the swirling water scoops out hollows and dents in the surface it travels across. New sediment then settles out of the water and fills the hollows in again. This pattern remains on the base or sole of the new bed, resulting in sole marks.

- Sole marks indicate the direction of current flow. At this site, the turbidity currents flowed from what is presently the south. As you stand facing the cliff, it's as if the current is swirling toward you.

Turbidite layers overturned by folding make it easy to view sole marks: (a) sole marks on an overturned layer; (b) a detailed view of sole marks.

## Related Outcrops

Evidence suggests that the sandstone and shale layers at this site are related to the Mainland sandstone, and that both formed in the same foreland basin (see site 5).

Rocks including those beside the lighthouse at Lobster Cove Head are part of a large block within the Rocky Harbour melange.

# Ahead of the Plow

## Disrupted Rock Layers in Melange, Rocky Harbour

Lobster Cove lighthouse stands in a peaceful setting. On a sunny day, the tidy green lawn and white buildings whisper that all is right with the world. But in the cliff face below this idyllic scene, rock layers are broken, distorted, and shifted, recording events that occurred during transport of the Humber Arm allochthon.

Lobster Cove Head, on which the lighthouse sits, is a raft of sedimentary layers preserved in and engulfed by the Rocky Harbour melange. The melange itself is a flaky, black shale best seen in the town it's named for.

As the Humber Arm allochthon pushed westward onto Laurentia, it slid along on the weak shale. What happened in the area around Rocky Harbour has been described by legendary Newfoundland geologist Hank Williams as the "snow plow effect." The rocks around the cove were subject to intense, sometimes chaotic movements as the allochthon shoved its way westward.

Lobster Cove Head and other sites nearby illustrate this effect on many different scales.

# Getting There

## Driving Directions

Follow Route 430 to Rocky Harbour. Main Street in Rocky Harbour leads to all three parking locations: (1) At the north end of Main Street, follow a side road (turn at N49.60426, W57.94725) to the parking area near the lighthouse. (2) Drive to the intersection of Main Street and Harbour Drive. (3) Follow Main Street south and west around the shoreline to the cemetery and fish plant.

## Where to Park

Parking Locations:

1. N49.60307, W57.95359 (lighthouse parking lot)

2. N49.59076, W57.91620 (hotel parking lot)

3. N49.58605, W57.93817 (gravel area by cemetery)

## Walking Directions

See individual outcrop descriptions for walking directions.

## Notes

Lobster Cove Head (location 1) lies within Gros Morne National Park – you must have a valid Parks Canada pass to hike the trails there. For details visit www.pc.gc.ca/grosmorne. No park pass is required to visit locations 2 and 3 in the community of Rocky Harbour.

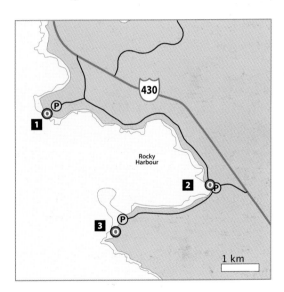

---

**1:50,000 Map**

Gros Morne 012H12

**Provincial Scenic Route**

Viking Trail

# On the Outcrop (1)

Sedimentary layers are fractured and disrupted at Lobster Cove Head. Signs of disruption include (a) broken beds and (b) a shear zone.

## Outcrop Location: N49.60204, W57.95566

In the parking lot, face the lighthouse and look to your left. You'll see a trail marked with a large boulder where it enters the woods. Follow this trail to a stairway (N49.60199, W57.95532) and the shore. Within 100 metres of the stairway in either direction, all the key rock types at Lobster Cove Head can be viewed, if you plan your trip for low tide.

Although the sequence is fractured and disrupted, the oldest rocks are generally found due west of the lighthouse, while younger ones occur south and east around the headland. The oldest rocks include distinctive ribbon limestones and a wide layer of conglomerate.

A younger group of dark shale layers alternate with thick (up to 1 metre across) light brown dolostone (a calcium-magnesium carbonate). These two rock types lie above a regional unconformity found throughout the Appalachians (see site 4).

The last rock type deposited at this site is a greenish grey sandstone that dominates the headland's north and south shores. The sand was eroded from the approaching allochthon and contains chromite, as the Mainland sandstone does (see site 5).

| 1000 | 900 | 800 | 700 | 600 | 500 |
|---|---|---|---|---|---|
| $Z_1$ | | $Z_2$ | | $Z_3$ | $\varepsilon$ |

# On the Outcrop (2)

Dark, flaky shale characterizes beach outcrops exposed at low tide in the town of Rocky Harbour.

## Outcrop Location: N49.59113, W57.91731

From the parking lot, walk across Main Street to the gazebo overlooking the harbour. As you face the water, to the right of the gazebo is a path to the shore. Walk down the path and continue north to visit the outcrop.

Like shadows, the mounds of shale at this site are dark against the pebbles and seaweed. The rock is black because it formed in deep water containing little oxygen but rich in organic carbon. Deformation squeezed the shale as the Humber Arm allochthon pushed westward onto Laurentia. Minerals aligned to form the thin layers along which the rock breaks so easily.

Rocky Harbour melange (detail).

Black shale is the only rock type in many parts of this site. However, in some outcrops on and around the beach, the flakes of shale are distorted around less deformed rock fragments.

500    400    300    200    100    0

€ O S D C P T J K Cz

# On the Outcrop (3)

Dark shale of the Rocky Harbour melange surrounds blocks of many sizes along the shore south of the fish plant in Rocky Harbour.

## Outcrop Location: N49.58423, W57.93970

Follow the footpath along the side of the cemetery. Just beyond the south end of the cemetery, follow the right-hand path fork down to the beach. The outcrop location noted above is about 100 metres farther south. Once you are on the shore, it is easy to explore other outcrops back toward the fish plant when tide conditions allow.

Along this section of the shoreline the outcrops contain numerous blocks of sedimentary rock. The blocks are like small-scale models of Lobster Cove Head because each one is surrounded by the black shale of the Rocky Harbour melange.

From this site on a clear day you can see Lobster Cove Head across the water. It's a good opportunity to picture the extent of the melange and the scale on which the "snow plow effect" operated.

1000          900          800          700          600          500

$Z_1$                          $Z_2$                    $Z_3$        €

# FYI

- The sediments at these three sites were deposited under similar conditions to those at Table Point and Mainland (sites 4 and 5) as the Iapetus ocean began to close. The approaching Humber Arm allochthon warped the edge of the continent as it pushed onto Laurentia, then travelled over the terrain to its present location, creating the melange.

- The shale of the Rocky Harbour melange (which began as a mud) still contained a lot of water as the Humber Arm allochthon moved westward over this area. The water helped reduce friction as the allochthon slid along. Hydrocarbons from the organic matter in the shale may also have helped reduce friction.

- Some geologists call a rock a "melange" only if the blocks within it are truly chaotic, with little or no trace of their original layering. Examples like those at the lighthouse and near the fish plant, where the layers can still be recognized, are sometimes called "broken formations" instead.

## Related Outcrops

The Rocky Harbour melange (yellow) lies beneath the Humber Arm allochthon and has been traced along the coast for 25 kilometres. It surrounds several kilometre-scale rafts of sediment (red) as well as innumerable smaller blocks.

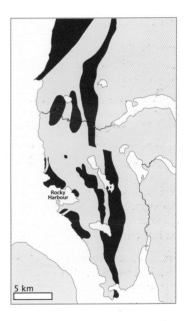

500    400    300    200    100    0

€  O  S  D  C  P  ?  J  K  Cz

A gravel road along the shore winds past outcrops of tectonic melange near L'Anse aux Meadows.

# Tectonic On-Ramp
## Melange at L'Anse aux Meadows

The Vikings who arrived at L'Anse aux Meadows more than 1,000 years ago had a lot in common with the Hare Bay allochthon: Both came from the east and made landfall at the tip of Newfoundland's Great Northern peninsula, leaving signs and traces for future generations to discover and wonder about.

Although the Vikings have vanished, the Hare Bay allochthon is still very much in place, the dominant geological feature between Hare and Pistolet bays. It's the northern counterpart to the Humber Arm allochthon, made up of thrust slices pushed onto Laurentia from the Iapetus ocean floor.

At L'Anse aux Meadows a tectonic melange – the rock unit along which the Hare Bay allochthon travelled – occurs along the shore. It's well exposed and easily accessed, with features typical of similar melanges found in many parts of Newfoundland.

These significant outcrops are just minutes from the UNESCO/National Historic Site and adjacent to Norstead, a replica Viking village.

# Getting There

## Driving Directions

On Route 436, from the entrance to the L'Anse aux Meadows National Historic Site continue northwest about 1 kilometre, passing Norstead Road. Then turn right (northeast) onto an unnamed road and follow it for about 300 metres along the shoreline.

## Where to Park

Parking Location: N51.60175, W55.52687

The parking area is beside a low building, near the shore.

## Walking Directions

From the parking area, a gravel road continues northeast along the shore toward the Norstead Viking Village. Cross onto the gravel beach at any convenient point and continue along the shore to the outcrop location.

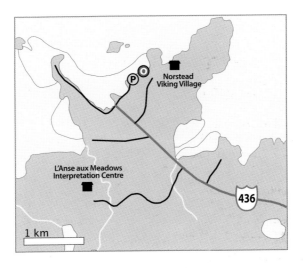

---

**1:50,000 Map**

Raleigh 002M12

**Provincial Scenic Route**

Viking Trail

# On the Outcrop

Rock layers of the melange at L'Anse aux Meadows tilt westward, roughly in the direction along which the allochthon moved. Blocks in the shale matrix take a variety of forms as shown below: (a) blocky; (b) rounded; or (c) flattened.

## Outcrop Location: N51.60234, W55.52530

By definition a mixture, typical melanges are recognized by a block-in-matrix fabric, in which easily deformed rock surrounds more resistant fragments.

At L'Anse aux Meadows, the matrix that surrounds the blocks was once a muddy sediment. As the Hare Bay allochthon travelled westward overhead, the matrix deformed, flowing and squeezing in response to the pressure and movement.

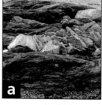

The blocks illustrate various states of modification – some are still quite fresh-looking, with clean, straight edges (photo a), while others are worn, broken, and rounded off (photo b). In a few cases the blocks, as well as the matrix, have been flattened and stretched (photo c).

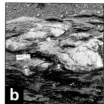

These differences reflect local variations in the amount of deformation, the amount of fluid in the rock while it deformed, and the strength of the block materials.

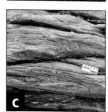

| 1000 | | 900 | | 800 | | 700 | | 600 | | 500 |
|---|---|---|---|---|---|---|---|---|---|---|
| | $Z_1$ | | | | $Z_2$ | | | | $Z_3$ | $\mathbb{C}$ |

# FYI

- Tectonic melanges occur in mountain belts the world over: for example, around the Pacific rim (California, Japan, New Zealand), as well as in parts of the Appalachians (including Newfoundland), Alps, and Himalayas.

- How do blocks get into a melange? Easily deformed sediment may squeeze into cracks in adjacent rocks, fracturing them and carrying the pieces along as movement continues. Or melange may originate as soft sediment into which fragments of rock layers have fallen from above.

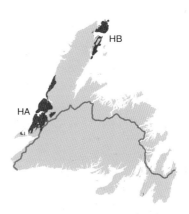

- Rock, mineral, and fossil ages suggest the Hare Bay allochthon (HB on map) was the first to be pushed onto Laurentia, then the Humber Arm allochthon (HA on map), then others farther south (for example, in Quebec). If so, the Iapetus ocean may have closed like a pair of scissors, north to south.

The uppermost slice of the Hare Bay allochthon includes ultramafic rocks like those of the Tablelands (see site 11). They occur, for example, on Mount Mer, seen here from Route 430 east of St. Anthony.

## Related Outcrops

Melanges are a common and significant feature of Newfoundland geology, particularly in the Humber and Dunnage zones. See sites 6, 7, and 9 for additional examples.

The Hare Bay allochthon itself includes thrust slices containing ocean-derived sediments, ocean crust, and underlying mantle.

500        400        300        200        100        0

€    O    S    D    C    P    Ŧ    J    K    Cz

On the beach and headland at Bottle Cove are several walking trails. The outcrops described here are along the shore beneath the hill.

# Misplaced Island

## Little Port Complex at Bottle Cove

As happens in the ocean basins of today, Iapetus ocean crust must have formed over many millions of years at mid-ocean ridges, above subduction zones, and in other tectonic settings (see Plate Tectonics, pages 13-15). Yet the Humber Arm allochthon preserves only small fragments of the once-wide Iapetus. So we wonder: "What kind?" "How old?"

At Bottle Cove – and elsewhere along the west coast of Newfoundland between Little Port and Bonne Bay – lies the Little Port complex. It's the oldest piece of ocean crust preserved in the Humber zone, about 20 million years older than Blow Me Down massif (see site 10) or the Tablelands (see site 11).

Chemical analyses of rocks from sites including Bottle Cove, Blow Me Down Provincial Park, and Little Harbour suggest the rocks formed in an island arc, which also makes this complex unique in the Humber zone.

# Getting There

## Driving Directions

Follow Route 450 for about 1.5 kilometres, from the entrance of Blow Me Down Provincial Park toward Lark Harbour. Turn north onto Little Port Road and follow it for a little more than 2 kilometres. At the fork, bear northeast onto Beacon Road and drive a further 200 metres to the parking area.

## Where to Park

Parking Location: N49.11342, W58.40658

There is a large gravel parking lot adjacent to Beacon Road, with signage for the boardwalk trail head.

## Walking Directions

From the parking area, follow the boardwalk to its termination and continue along the sandy track to the beach. Walk along the beach and then among boulders to the outcrop location.

## Notes

Follow Beacon Road about 0.5 kilometres farther – around the cove and up the hill – for alternative parking and access to additional walking trails.

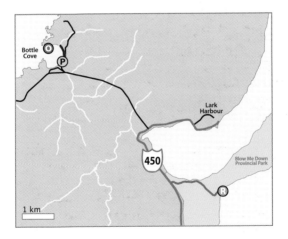

---

**1:50,000 Map**

Bay of Islands 012G01

**Provincial Scenic Route**

Captain Cook's Trail

# On the Outcrop

Along the shore at Bottle Cove are island-arc volcanic rocks as well as evidence of their transport: (a) shale melange; (b) a melange of chaotic fragments; (c) a block of pillows and pillow breccia; and (d) well-preserved pillow lava.

**Outcrop Location: N49.11623, W58.40989**

As you walk along the shore from the end of the boardwalk toward the outcrop, you'll encounter a series of related rock types. First to appear is a melange with a shale matrix so fragmented and flaky it barely holds together on the hillside. Even the recognizable boulders it contains are fractured.

The melange changes character farther along the shore. The matrix is greenish because it includes altered ocean crust. There is less matrix, and the fragments are a chaotic mixture of rock types.

Next to the melange is an area of larger blocks 2-4 metres across. Their interiors are well preserved, but they're surrounded by melange. One of these large blocks contains altered pillow lava and pillow breccias.

Finally, at the specified outcrop location is a similar large block containing well-preserved whole pillows. Like the pillow lava at Mine Brook (see site 10), it formed when the hot lava came into contact with colder sea water.

| 1000 | 900 | 800 | 700 | 600 | 500 |

$Z_1$     $Z_2$     $Z_3$     €

# FYI

- The Little Port complex, along with the Sleepy Cove lava and Twillingate granite in the Dunnage zone (see sites 13 and 14), are two of the oldest tracts of Iapetus ocean crust in Newfoundland.

- The Little Port complex lies west of the Bay of Islands complex (which includes Blow Me Down mountain and the Tablelands; see sites 10 and 11) between the Serpentine river and Bonne Bay. As shown on the topographic map below, the boundary between the two (Little Port complex in red; Bay of Islands complex in yellow) usually takes the form of a broad valley due to erosion of the weaker melange layer that separates them. As you travel west toward York Harbour along Route 450, you'll see the striking break in topography between these oceanic rocks of different ages.

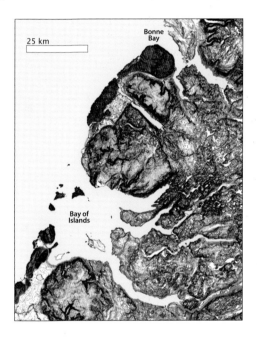

## Related Outcrops

On Route 431, about 3.2 kilometres west of the Gros Morne Discovery Centre, a pull-off (N49.48054, W57.96144) provides a scenic view of the Tablelands. Across the highway is a road cut containing pillow lava as well as deep ocean sediments belonging to the Little Port complex.

In Blow Me Down Provincial Park, granitic portions of the Little Port complex are well exposed. The Governor's Stairway (N49.09085, W58.36285) climbs right through the rock formation.

Because the complex was on the western (leading) edge of the Humber Arm allochthon, its rock formations tend to be disrupted, fragmented, and cut by melange (also see site 7). This effect is particularly evident in the cliff faces at Little Port harbour (N49.10635, W58.42158).

| 500 | 400 | 300 | 200 | 100 | 0 |
|---|---|---|---|---|---|
| Є | O | S | D | C | P | T | J | K | Cz | |

Pillow lava frames the entrance of an old mine near York Harbour, Humber Arm South. Blow Me Down mountain is in the background.

# Underwater Eruptions
## Pillow Lava at Mine Brook

From the rest stop and trail head at Mine Brook, the crags of Blow Me Down mountain loom in the east. The view is appropriate because this site and the mountain in the distance are both part of a single ophiolite sequence detached from the Iapetus ocean and thrust over the Humber zone during the middle of the Ordovician period.

The rocks here formed on the ocean floor itself. To the east and, to a lesser extent in the west, are rocks from deeper levels of the ocean crust and even mantle.

The formation of pillow lavas is an underwater process, messy and dramatic. When hot lava emerges into water, it forms a rounded shape – a pillow – that quickly crusts over as its outer surface cools. More lava breaks the crust open and flows out to form another pillow, and so on. Heated sea water circulates through the mounded pile of pillow forms, creating a rich brine that may deposit mineral veins.

The pillows are well preserved at this site – you can practically hear the lava sizzle.

# Getting There

### Driving Directions

Follow Route 450 along the south side of Humber Arm. About 5 kilometres east of York Harbour, turn onto a gravel drive that angles up the hill on the south side of the road.

### Where to Park

Parking Location: N49.06187, W58.30434

The parking area is about 150 metres from the highway at the top of the drive. This is a large gravel area and rest stop.

### Walking Directions

The outcrop location is beside the entrance to an old mine in the hillside on the southern edge of the parking lot, near the top of the gravel drive.

### Notes

While not part of a provincial or national park, this site serves as a trail head and rest area for the Copper Mine to Cape Trail, part of the International Appalachian Trail.

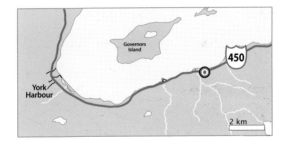

**1:50,000 Map**

Bay of Islands 012G01

**Provincial Scenic Route**

Captain Cook's Trail

# On the Outcrop

Well-preserved pillow lava bulges from the mine entrance. Its bumpy surface texture formed during rapid cooling.

**Outcrop Location: N49.06176, W58.30407**

The ground drops off on the north side of the parking/rest area, while on the south side it rises steeply into the adjacent hill. In the steep south wall, framing the entrance to the mine, the bulbous, reddish forms of the pillows are obvious. Notice how the round shapes are moulded to accommodate one another – each pillow was soft when it first erupted as hot lava into the Iapetus ocean.

As would be expected in an ophiolite sequence, these pillows are made of basalt. Their reddish colour is due to the iron oxide mineral hematite, probably formed by alteration of the rock as it interacted with sea water.

The surface of several individual pillows at this site is covered with small bumps of uniform size. Geologists call this surface texture spherulitic. It forms when hot lava is supercooled by contact with cold sea water. The spherules are formed of microscopic bundles of crystals that gather at a single point, a bit like a dandelion gone to seed.

| 1000 | 900 | 800 | 700 | 600 | 500 |
|---|---|---|---|---|---|
| | Z₁ | | Z₂ | Z₃ | € |

# FYI

- The mine shaft at this site was created in the 1960s in hopes of finding a copper deposit similar to one previously mined at York Harbour. However, because the lavas are at the base of the thrust slice (see diagram), the shaft quickly passed through into shale melanges, and the mine was never developed.

- The diagram below shows the position of the Mine Brook pillow lava in relation to other parts of the ophiolite preserved in Humber Arm South. The sequence from mantle to oceanic crust to ocean floor lavas was folded into a syncline, then cut off along the bottom as it was pushed onto the continent.

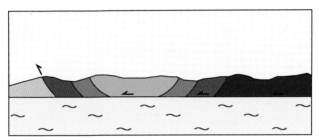

A schematic slice cutting east to west down through Blow Me Down mountain and related parts of the ophiolite sequence in Humber Arm South. The diagram represents an area about 1.5 kilometres deep by about 15 kilometres across (not to scale). Pillow lava is lightest red; darker shades of red indicate successively deeper layers of the ocean crust and mantle. Light grey is melange; dark grey is the Laurentian continental margin.

## Related Outcrops

Like all ocean crust rocks in the Humber zone, these originated in the Dunnage zone to the east. The Dunnage zone contains many pillow lavas. See sites 13 and 16 for additional examples.

## Exploring Further

US National Oceanic and Atmospheric Administration website, oceanexplorer. noaa.gov. Search "pillow lava" to see a video of pillow lava being formed underwater.

Yellow-weathering peridotite creates a stark landscape in Winterhouse Brook canyon, Gros Morne National Park.

# Earth – or Mars?

## Earth's Mantle Exposed in the Tablelands

Driving by the Tablelands is a strange experience. On one side of the road, the hills are forested and green. The other side is like the bare surface of an alien planet, strewn with yellowish brown rubble.

The Tablelands looks out of place, and it is. This uppermost layer in the Humber Arm allochthon is a slab of the Earth's mantle that originated several kilometres below the Iapetus ocean floor. Thanks to the scouring action of ice-age glaciers, the Tablelands provides some of the most extensive, pristine outcrops of mantle rock in the world.

The rock – peridotite – is chemically unstable in the cool, damp environment of the Earth's surface. Water seeping through cracks in the Tablelands is slowly changing peridotite to serpentinite, in turn altering the water chemistry. It's a process that was probably widespread during the early history of the Earth and may have played a role in the origin of life.

American and Canadian space agencies have designated the Tablelands as an official Mars analog because peridotite and serpentinite have also been found on that planet. Experiments that check for signs of life in such terrains are developed and tested here in preparation for future Mars missions.

At Winterhouse Brook, you can see some of the features that have attracted space researchers' attention.

# Getting There

## Driving Directions

Follow Route 431 in Gros Morne National Park. Between Woody Point and Trout River, about 4.5 kilometres west of the park's Discovery Centre, turn into a gravel drive on the south side of the road. Drive about 200 metres to the parking location.

## Where to Park

Parking Location: N49.47813, W57.97399

Follow the gravel drive on the south side of Route 431 to the parking lot.

## Walking Directions

From the parking lot, follow the gravel track and boardwalk to Winterhouse Brook. The boardwalk ends at a viewing platform near a small waterfall in the brook.

## Notes

There is no cost specific to the Tablelands site itself, but to hike this trail you must have a valid Parks Canada pass (see www.pc.gc.ca/grosmorne for details). Guided tours are available but not required.

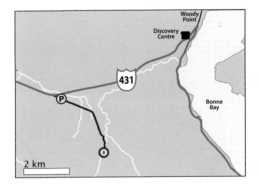

---

**1:50,000 Map**

Lomond 012H05

**Provincial Scenic Route**

Viking Trail

# On the Outcrop

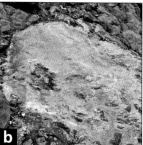

Near the waterfall at the observation deck along Winterhouse Brook, peridotite outcrops show signs of serpentinization: (a) seeps, and below them, (b) travertine deposits.

## Outcrop Location: N49.46602, W57.95712

Water from ordinary rain or snow soaks into the Tablelands through cracks and crevices, but by the time that water emerges from the ground in the form of seeps, it's as alkaline as household bleach, lye, or ammonia. The seeps also contain tiny bubbles of hydrogen and methane gas, as well as dissolved minerals such as calcium. These changes all happen as the water converts peridotite to serpentinite.

Near the observation deck, you can locate a few of these seeps (photo a). Where they emerge, travertine is deposited. It's a form of limestone that builds up in thin layers, so look for smooth, pale surfaces among the brown peridotite (photo b). In some places the travertine has cemented the stream gravel together to form a sort of conglomerate mound.

Nearby, above the travertine, look for a small pool of water, where the seep may have filled a depression in the rock. These small pools are the subject of intensive research. Even though the water is so alkaline and anoxic, it contains microbial life forms. How do they survive? What do they live on? The answers to these questions could help us understand the origins of life.

| 1000 | 900 | 800 | 700 | 600 | 500 |
|---|---|---|---|---|---|
| $Z_1$ | | $Z_2$ | | $Z_3$ | $\mathbb{C}$ |

# FYI

- When serpentinite forms, it takes up more room than the peridotite did. So the conversion process stresses the rock, creating more cracks and allowing more water into the rock, accelerating the process further.

- Water in the seep pools has a pH of 12 on a scale of 0-14 (household bleach is pH10).

- Microbes that survive in harsh conditions such as the seep pools are known as extremophiles. They may be similar to life forms that could survive in extraterrestrial environments, for example on the surface of Mars.

- Gros Morne National Park's Trout River Pond boat tour provides a detailed look at the Tablelands ophiolite including views of the Moho, or boundary between ocean crust and mantle. Visit www.pc.gc.ca/grosmorne or the Discovery Centre on Route 431 for details.

## Related Outcrops

Ultramafic rocks similar to those in the Tablelands are found in several locations (shaded red in the map above). All originated below the Iapetus ocean floor:

- White Hills peridotite north of Hare Bay on the Great Northern peninsula.

- North Arm massif, north shore, Bay of Islands.

- Blow Me Down massif and the Lewis Hills, Humber Arm South.

- Smaller bodies of ultramafic rock along the eastern and western boundaries of the Dunnage zone. See sites 18 and 27 for examples of these.

The stream bed of Blanche Brook in Stephenville is wide and boulder-strewn, ideal for exploring when water levels are low.

# Upland Pioneers
## Fossil Trees in Blanche Brook, Stephenville

The hillsides of Newfoundland were not always forested as they are now. Trees first appeared during the Devonian period but were restricted to wet, low-lying areas. So when did evolution finally "green" the broader landscape? Blanche Brook in Stephenville offers evidence of the first trees to grow in a hilly upland environment of thin, well-drained soils.

The fossil trees in Blanche Brook were discovered more than 135 years ago. The Newfoundland Geological Survey's "Report for 1873" describes that field season's curious finding: "In the valley of the Rivière Blanche ... one surface ... is strewed over with trunks, limbs, and branches of Carboniferous trees; so that, as seen from a little distance, the appearance of the ground reminds one of a windfall or drift of modern wood upon a beach of sand."

In 2004 a graduate student at Memorial University made an intensive study of these fossil trees, revealing details about their form, great size, and importance as evolutionary pioneers.

The town of Stephenville, in partnership with the International Appalachian Trail of Newfoundland and Labrador, has constructed a trail that makes it easy to visit parts of this significant fossil record.

# Getting There

## Driving Directions

Follow Route 460 to Stephenville. On that route, about 2 kilometres northeast of Kippens Road, look for a motel on the south side of the road. Nearby on the north side, immediately east of the bridge over Blanche Brook, look for a gravel parking area with signs for the fossil forest. This is the trail head.

## Where to Park

Parking Location: N48.56247, W58.58166

Park in the gravel lot beside the bridge, on the north side of Route 460 where it crosses Blanche Brook.

## Walking Directions

From the trail head in the corner of the parking lot, follow the path along the east bank of Blanche Brook and then into a wooded area. About 1 kilometre from the trail head, an obvious clearing along the bank allows a short climb down into the rocky stream bed when water levels are low.

## Notes

This path is one of a set of walking trails maintained by the town of Stephenville; the fossil site is protected by law. It is illegal to disturb, damage, or remove any fossil material from this site.

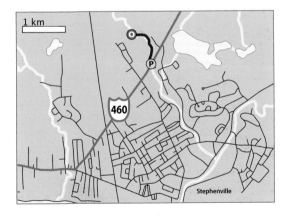

---

**1:50,000 Map**

Stephenville 012B10

**Provincial Scenic Route**

French Ancestors Route

# On the Outcrop

A carbonized fossil stump is surrounded by sandstone in Blanche Brook, Stephenville.

As you step into the stream bed at this site, you'll nearly set your foot right on a group of fossil tree limbs. They're preserved in a grey conglomerate that was deposited in the channel of a small, winding stream. Elsewhere in the stream bed, you'll see sandstones, too – formed as finer sediment accumulated on an adjacent floodplain.

One of the trunks here has been worn away, revealing a pattern of concentric layers in the fossil wood. These are not thought to be annual growth rings but may represent periodic growth interruptions of some kind.

A little farther upstream, a few metres from the bank, a well-preserved stump is plainly visible, especially when water levels are low. Its black surface is carbonized (part of the fossilization process), making it easy to see the flared base and roots. Analysis of all the fossil roots in Blanche Brook suggests the trees grew in shallow soil.

Water flow in these small Carboniferous streams was probably not powerful, and the tree roots are mostly well preserved. This suggests the trees were not carried far. Most likely they grew nearby and fell when the stream undercut the bank.

1000     900     800     700     600     500

$Z_1$         $Z_2$         $Z_3$     €

# FYI

- More than 200 tree fragments have been identified in Blanche Brook.

- The internal cell structure of the fossilized wood confirms the trees at Blanche Brook are all the same species.

- Correlations using other plant and pollen fossils at the site were used to determine the age of the rocks in Blanche Brook.

- The trees belong to an extinct order of plants, Cordaitales. Its closest living relatives are primitive cone-bearing trees of the southern hemisphere (for example, the bunya tree of Australia and the monkey puzzle tree of Chile).

A reconstruction of the whole tree (left) based on the fossil tree parts in Blanche Brook. Growing as high as 48.5 metres, they were probably among the tallest trees living at the time (Bashforth, 2005).

## Related Outcrops

Most rocks of Carboniferous age in Newfoundland lie above the Humber zone in a narrow region known as St. George's Basin, and around Deer Lake (both shaded red in the map at right).

In Canada, other deposits of similar large Pennsylvanian trees are found in Nova Scotia and New Brunswick.

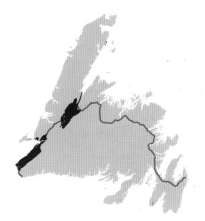

## Exploring Further

Bashforth, A.R. *Late Carboniferous (Bolsovian) Macroflora from the Barachois Group, Bay St. George Basin, Southwestern Newfoundland, Canada.* Palaeontographica Canadiana, no. 24. St. John's: Canadian Society of Petroleum Geologists and Geological Association of Canada, 2005 (123 pages, 20 plates; available from the Geological Association of Canada).

| 500 | 400 | 300 | 200 | 100 | 0 |
|-----|-----|-----|-----|-----|---|
| Є O | S D | C P | Ŧ J | K | Cz |

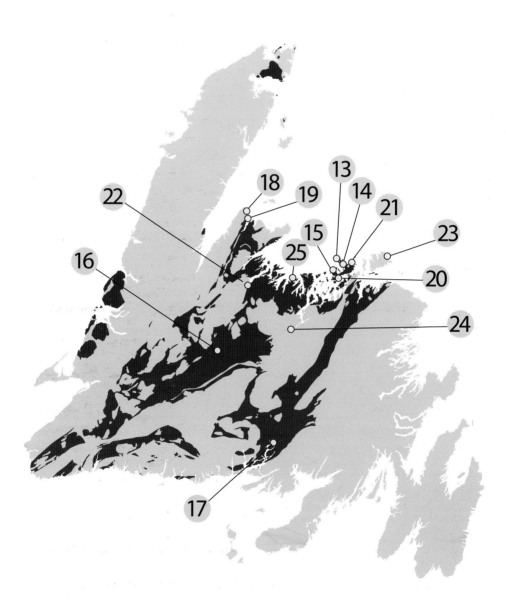

# DUNNAGE

## Zone at a Glance

### Boundaries

West: Baie Verte-Brompton Line
East: GRUB Line

### Origin

Iapetus ocean floor and islands

### Characteristic Features

Iapetus ocean: Sediment–crust–mantle;
  sea-floor mineralization
Closure–collision: Intrusions,
  volcanoes; river sediment

## In the Dunnage zone, you can ...

| | | |
|---|---|---|
| 13 | Sleepy Cove | Find early traces of an ancient ocean floor. |
| 14 | Little Harbour | View the root of an ocean island. |
| 15 | Moreton's Harbour | See how ocean crust formed millions of years ago. |
| 16 | Millertown | Track down a mineral deposit. |
| 17 | Trout Hole Falls | Stand on sediments from the Iapetus ocean floor. |
| 18 | Fleur de Lys | Learn how bits of Earth's mantle come in handy. |
| 19 | Coachman's Cove | Visit the boundary between the Humber and Dunnage zones. |
| 20 | Summerford | Climb to a panorama of the spectacular Dunnage melange. |
| 21 | Pike's Arm | Explore rock debris that filled in a vanishing ocean basin. |
| 22 | Goodyear's Cove | Walk into the realm of an explosive volcanic eruption. |
| 23 | Tilting | Track down rocks from the bottom of a magma chamber. |
| 24 | Bishop's Falls | Examine ancient river sediments along the banks of the Exploits river. |
| 25 | Leading Tickles | Discover signs of the Atlantic ocean's beginnings. |

The Dunnage zone did not exist at all prior to the opening of the Iapetus ocean. That's because the Dunnage zone is made of leftover bits and pieces of that ancient ocean realm. It's a veritable museum of oceanic rock types, including ophiolites, volcanic arcs, and sedimentary rocks typical of deep ocean basins.

The history of the Iapetus ocean is complex. Although it is talked about as a single ocean, geologists now think a series of seaways and ocean basins opened and closed at different times between 540 and 430 million years ago. Present-day analogies may include the western Pacific ocean or Caribbean sea.

Moreton's Harbour (see site 15).

The oldest remnants from the Iapetus ocean found so far are about 510 million years old (sites 13 and 14). Ocean crust continued to form until about 460 million years ago (sites 15 and 16). Sedimentary rock continued to accumulate in the ocean basin even as it closed. The formation of sedimentary rock continued longer on the eastern side, the side nearer to Gondwana (site 17).

Evidence of one of the early stages of ocean closure can be found on the Baie Verte peninsula (sites 18 and 19). There, rocks equivalent to those in the Humber Arm allochthon are preserved in a steeply dipping zone of deformation, the Baie Verte-Brompton Line.

Wild Bight (see site 15).

Turpin's Trail, Tilting (see site 23).

Subduction of the mid-ocean ridge about 470 million years ago (site 20) was the beginning of the end for the Iapetus ocean. By 440 million years ago, parts of Gondwana and Laurentia were in contact along the Red Indian Line (site 21).

The last remaining basin in the Iapetus ocean finally closed along the Dog Bay Line about 430 million years ago. A complex continental collision began to squeeze and wrench the crust. Molten rock formed at deep levels, then rose through areas weakened by distortion. This resulted in volcanic eruptions (site 22) and magmatic intrusions (site 23) in many parts of the Dunnage zone.

During the continental collision, Laurentia was pushed up over Gondwana, so there was high ground west and north of the Red Indian Line. As erosion wore down the landscape, rivers deposited sediment in lower lying areas east and south of the collision zone. Red sandstones typical of these river deposits are common in parts of central Newfoundland (site 24).

Coachman's Cove (see site 19).

Formation of the Appalachian mountains was part of a larger process that created the supercontinent Pangaea. After a long period of quiet, the restless Earth began to reorganize. As Pangaea broke up to form the Atlantic ocean, cracks in the Earth's crust were filled with molten rock, forming dykes that can be seen in parts of the Dunnage zone (site 25).

Cliffs of pillow lava encircle Sleepy Cove on North Twillingate Island.

# Widening Ocean
## Pillow Lava at Sleepy Cove

In 1910 Sleepy Cove was far from sleepy. Obediah Hodder, of nearby Crow Head, had purchased the finest mining equipment money could buy and launched the Great Northern Copper Company Limited. Two industrial buildings loomed in the tiny cove; steam-powered machinery chuffed, clattered, and clanged to fill a waiting ship with copper ore.

But Obediah was a victim of his own unfounded enthusiasm. There was very little copper at the site, and by 1917 he had abandoned his operations.

Today, the cove is sleepy once again. Relicts of the mining misadventure are brightly painted red and yellow. Picnic tables dot the area, which is part of a system of walking trails on North Twillingate Island.

Geologists have a renewed interest in the site unrelated to its mining history. The pillow lava in the cove is among the oldest volcanic rocks in the Dunnage zone – that is, the oldest remnants of Iapetus – probably formed as part of an island arc.

# Getting There

## Driving Directions

Follow Route 340, which takes a 90-degree turn in the town of Crow Head on North Twillingate Island and heads north toward the lighthouse at Long Point. About 0.9 kilometres north of the turn, there is a gravel road on the west side. Follow it toward the shore. At the turn-off, you may see a sign for walking trails, including Nanny's Hole, Lower Head, and Sleepy Cove.

## Where to Park

Parking Location: N49.68367, W54.80657

At the shore there is a large open area encircled by the gravel/dirt road. Park near the north end for quicker access to the outcrop.

## Walking Directions

From the parking area, walk north toward the cove and follow one of the paths leading onto the sandy beach.

## Notes

Sleepy Cove is part of the Twillingate Islands Coastal Trails System. Interpretive panels at this site and other key locations display a map of the trails. The outcrops are visible at high tide but the beach may not be accessible in all conditions.

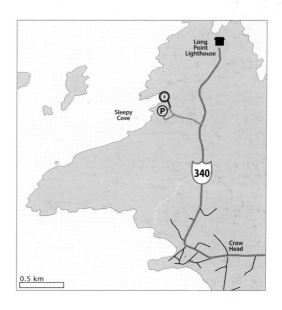

---

**1:50,000 Map**

Twillingate 002E10

**Provincial Scenic Route**

Kittiwake Coast: Road to the Isles

# On the Outcrop

Pillow lava is exposed in the cliff face at Sleepy Cove, near Crow Head, North Twillingate Island. The light grey object in the foreground is a remnant of historic mining operations.

**Outcrop Location: N49.68449, W54.80656**

The cliff face that shelters Sleepy Cove is composed mostly of a dark green rock called basalt. This volcanic rock erupted underwater to form bulbous shapes called pillow lava.

The pillow lava at Sleepy Cove is a little worse for wear. The greenish colour is due to metamorphism, which caused the original igneous crystals in the rock to be replaced with chlorite and other lower-temperature minerals.

The pillows have been deformed by later events, too, so they are somewhat elongated. Even so, in the middle part of the cliff you can still clearly see the rounded shapes about 50 centimetres across.

To your right as you face the cliff, on the south end of the cove, the pillow lava has been caught up in movement along a later fault. The shapes have been flattened and obliterated in a zone parallel to the fault.

In the cliff face at the cove you will also see two dykes, narrow walls of once-molten rock that intruded the pillow lava. One of the dykes is light grey and the other is dark brown. They were intruded into the pillow lava during unknown later events.

| 1000 | 900 | 800 | 700 | 600 | 500 |
|------|-----|-----|-----|-----|-----|
| | Z₁ | | Z₂ | Z₃ | € |

# FYI

- The pillow lava here is of similar age to pillow lava in the Little Port complex in the Humber zone (see site 9).

- A few other areas of the Dunnage zone preserve volcanic rocks about 500 million years old. Examples include the Lushes Bight volcanic rocks on the northern edge of Pilley's Island and a narrow belt of volcanic rocks around Lake Ambrose south of Millertown.

## Related Outcrops

All the outcrops on North Twillingate Island are volcanic rocks related to those at Sleepy Cove. Any of the walking trails in that area will bring you in contact with them. You can see other examples of the pillow lava from the viewing platform near the lighthouse at Long Point, about 0.6 kilometres north of Sleepy Cove.

Pillow lava of the Sleepy Cove Group is visible from the viewing platform at Long Point, North Twillingate Island.

The Twillingate granite (see site 14) intrudes the pillow lava of the Twillingate islands. The two rock types may represent different parts of a single island arc formed early in the history of the Iapetus ocean.

## Exploring Further

Martin, Wendy. *Once Upon a Mine: Story of Pre-Confederation Mines on the Island of Newfoundland*. Special Volume no. 26. Quebec: Canadian Institute of Mining and Metallurgy, 1983. (Pages 26-27 of the book provide details of Obediah Hodder's enterprise.)

500     400     300     200     100     0

Є   O   S   D   C   P   Ꝑ   J   K   Cz

Outcrops of Twillingate granite dot the landscape in Little Harbour, South Twillingate Island.

# Island Root

## Twillingate Granite at Little Harbour

Today: Twillingate is an island. About 510 million years ago: Twillingate was an island. Mind you, it was an island of a very different sort …

Out in the wide Iapetus ocean, a subduction zone formed late in the Cambrian period. Volcanoes spilled out pillow lava underwater, piling it up to create a mountain in tropical seas just south of the equator. It may have been like part of the Antilles islands in the Caribbean of today.

Eventually the crust beneath the island began to melt. Granitic rock formed and intruded the pillow lava above. Eventually subduction stopped in this part of the ocean, leaving the island rocks to cool and solidify.

What you see on Twillingate island today are the remnants of that Cambrian island. The Twillingate granite at Little Harbour formed at its root and intruded into the pillow lava of Sleepy Cove and related areas (see site 13).

# Getting There

## Driving Directions

Follow Route 340 to the area a few kilometres south of the town of Twillingate. Watch for signs to Little Harbour, and turn east onto Little Harbour Road, which loops from Route 340 out to the coast and back. Follow the road to the west side of Little Harbour, watching for a picnic area east of the road.

## Where to Park

Parking Location: N49.63121, W54.71035

This is a gravel parking area beside a wooden platform and picnic table.

## Walking Directions

The outcrop is best exposed just beyond the hummocky uneven ground beside the parking area. Following any of several well-worn footpaths, make your way toward the shoreline.

## Notes

The nearby Lower Little Harbour to Jones Cove walking trail provides more opportunities to see the Twillingate granite in this area. You can visit the scenic Natural Arch along the way.

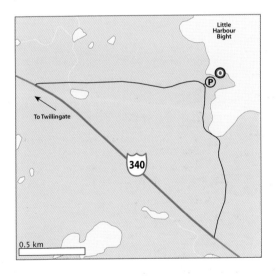

---

**1:50,000 Map**

Twillingate 002E10

**Provincial Scenic Route**

Kittiwake Coast: Road to the Isles

# On the Outcrop

Darker xenoliths are scattered through the light-coloured granitic rock at Little Harbour, South Twillingate Island.

**Outcrop Location: N49.63152, W54.70980**

The Twillingate granite at this site is made of medium-sized grains of quartz and feldspar with a small amount of dark minerals such as biotite. Because the granite has been deformed, the rock has a subtle fabric – in certain angles of sunlight, the prominent white quartz grains appear to be aligned in parallel layers.

In this outcrop there are many xenoliths of darker rock called amphibolite. Some of the xenoliths are long and narrow with straight, parallel sides. That suggests they are remnants of dykes that intruded the granite, perhaps before it was completely solid. Other xenoliths are less regular in shape, giving the impression they are older rocks swept up in the still-molten granite as it moved through the crust. In other words, the xenoliths don't have a clear-cut age relationship to the granite.

Some geologists have suggested that both the granite and the amphibolite formed as part of a single ongoing process that occurred deep in the crust.

1000   900   800   700   600   500

Z₁   Z₂   Z₃   €

# FYI

- Usually referred to as Twillingate granite, the light-coloured rock so common on South Twillingate Island is not a true granite at all, because it lacks potassium feldspar. More precisely, it's a type of tonalite called trondhjemite.

- A few other areas of the Dunnage zone preserve rocks about 500 million years old. Examples include the Lushes Bight volcanic rocks on the northern edge of Pilley's Island and a narrow belt of volcanic rocks around Lake Ambrose south of Millertown.

- Granitic rocks of the Little Port complex in the Humber zone (see site 9) have similar age and chemistry to the rocks at this site.

## Related Outcrops

The Twillingate granite intrudes pillow lavas like those at Sleepy Cove (see site 13). The two rock types represent different parts of a single island arc formed early in the history of the Iapetus ocean. They are surrounded by younger rocks.

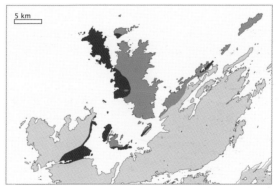

Light blue areas are Twillingate granite; dark blue areas are Sleepy Cove volcanic rocks; grey areas are younger rocks.

## Exploring Further

Twillingate Island Tourism Association website, www.twillingatetourism.ca. Follow links to download a map of the hiking trail from Lower Little Harbour to Jones Cove.

500     400     300     200     100     0

€   O   S   D   C   P   Tr   J   K   Cz

Lava, volcanic ash, and related intrusions from the Iapetus ocean floor frame a small cove on the east side of Moreton's Harbour, New World Island.

# Eruptions and Intrusions
## Ocean Floor Formation at Moreton's Harbour

The fame of Moreton's Harbour is secured forever by its mention in the Newfoundland song, "I'se the B'y." The song's enduring popularity reminds us of the strong ties that bind the region through history and folklore. But there is another common thread in a shared geologic past.

The Dunnage zone contains many remnants of the Iapetus ocean. There are literally thousands of outcrops in central Newfoundland illustrating processes that took place on and below the ocean floor.

Each outcrop is like a single frame clipped and selected by chance from a full-length movie. For decades, geologists have worked to piece the story together. At Moreton's Harbour you can view a "scene" in which ocean crust was created by a combination of volcanic eruptions and related igneous intrusions.

The scene took place during the Ordovician period, in a sea near the margin of the continent of Laurentia. Plate tectonic forces were pulling the existing ocean crust apart and new molten material was rising from below. Watch closely now …

# Getting There

### Driving Directions

On New World Island, follow Route 345 in Moreton's Harbour to the intersection with an unpaved road on the east side of the harbour. Follow the unpaved road about 300 metres to the area near the large satellite dish installation.

The second parking location farther east is for Wild Bight. See Related Outcrops.

### Where to Park

Parking Location: N49.57649, W54.85739

Park off the road, being careful not to obstruct harbour activities.

### Walking Directions

Near the satellite dish installation, you will find a vehicle track. Follow it until the footpath forks right from it, then follow the footpath. Walk over the higher ground into a smaller cove, then follow that shore around to the outcrop location.

### Notes

Moreton's Harbour is a working harbour, and the walking trail passes near private properties. While the outcrop is accessible at high tide, it is much easier to examine when low tide exposes a sandy area around the shore in front of it.

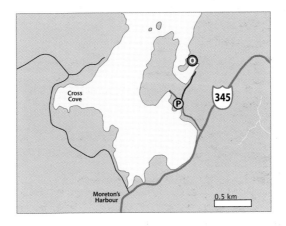

**1:50,000 Map**

Twillingate 002E10

**Provincial Scenic Route**

Kittiwake Coast: Road to the Isles

# On the Outcrop

Mottled volcanic ash (a) is cut by a smooth diabase dyke (b) at Moreton's Harbour, New World Island. In each photo, the contrasting rock type appears at the far left.

## Outcrop Location: N49.57931, W54.85568

The outcrop begins at the north end of a small beach and continues for about 30 metres along the shore. There are two main kinds of rock at the site: Some were erupted as lavas or ash and some were intruded in the form of dykes.

Both the lava and the ash are grey to greenish grey. In some places the lava is vesicular; in other words, it contains small voids that were originally bubbles of gas in the lava. In other places you can see a "splat" texture where blobs of lava fell and cooled. The volcanic ash contains fragments of rock in a finer matrix. It has a rough appearance and mottled colour.

Intrusive diabase dykes can be seen cutting across the erupted lava and ash. Compared to the rough or mottled surface of the lava and ash, the dykes have a smooth, even texture. In many places they have weathered to a soft brown colour.

1000    900    800    700    600    500

$Z_1$    $Z_2$    $Z_3$    $\epsilon$

# FYI

- New ocean crust is generated at a spreading centre where older ocean crust is pulled apart to form a narrow crack. Molten rock (magma) from hot regions below flows through the crack to erupt at the surface. Magma left inside the crack cools to become a solid dyke. Then new cracks form and the process continues.

- Mid-ocean ridges are the largest, but not the only, locations where spreading occurs. It's more likely the rocks at Moreton's Harbour formed above a subduction zone or in a small ocean basin behind an island arc.

- The rocks around Moreton's Harbour are part of a larger segment of Ordovician ocean floor that extends from New World Island westward to the Fortune Harbour peninsula. They may be part of an extensive chain of similar rocks called the Annieopscotch ophiolite belt.

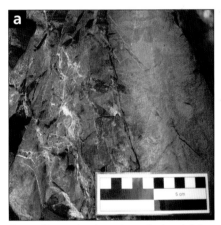 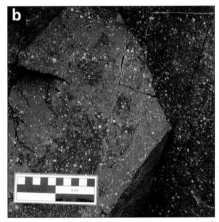

Ocean-floor volcanic rocks and related dykes can be seen around the shore at Wild Bight, New World Island: (a) a brown-weathering diabase dyke cutting veined volcanic rocks; (b) large feldspar crystals in dykes with a speckled appearance.

## Related Outcrops

For a related outcrop, continue east along Route 345 for 2 kilometres. Turn north and follow a short gravel road to a picnic area. Walk east from the picnic area to the small cove known as Wild Bight. Along the shore you can see rough, veined volcanic rock cut by smooth grey dykes. Some of the dykes contain visible feldspar crystals (phenocrysts).

Volcanic rock, mainly in the form of pillow breccia, outcrops along the shore of Red Indian Lake in Millertown.

# Circulating Minerals
## Mineralized Ocean Crust at Millertown

Red Indian Lake is rich in stories of the distant past. It featured prominently in the life of Newfoundland's Beothuk communities. They wintered on its shores, migrating down the Exploits river each year to spend the warmer months along Notre Dame Bay.

Red Indian Lake is also significant for the geological history of Newfoundland. A major tectonic boundary called the Red Indian Line runs along the length of the lake. Today that line is all that separates two once-distant regions. Iapetus ocean rocks that formed near Laurentia lie to the west of the line. Those formed near Gondwana lie to the east.

Rocks of the Iapetus ocean are rich in minerals along the Red Indian Line. Today, mine operations, the revitalization of older mines, and mineral exploration activities figure in regional strategic plans and in local news. On the shores of Red Indian Lake in Millertown you can view examples of mineralized volcanic rock from the Iapetus ocean.

Note: The Red Indian Line jogs a few kilometres eastward at the head of the Exploits, so Millertown itself lies on the Laurentian side.

# Getting There

## Driving Directions

From Route 370 in Buchans Junction, take Route 370-11 to Millertown. In Millertown, follow Lakeview, Beothuk, or other local streets from the highway down to the lakefront. Along the lake, you may see a sign for the Iron Wheel, an item of interest to the area's industrial history.

## Where to Park

Parking Location: N48.81173, W56.54547

This is a small gravel parking area adjacent to the trail for the Iron Wheel and lakeshore. The location is southwest of the Lewis Miller Room Museum on Beothuk Road.

## Walking Directions

Follow the trail through grass and brush to the lake, then walk southwest along the cobble shoreline. The main outcrop is just past the old mill.

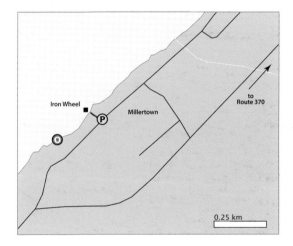

**1:50,000 Map**
Buchans 012A15

**Provincial Scenic Route**
Exploits Valley

# On the Outcrop

Sulphide mineralization appears as yellow areas and rusty staining in an outcrop of pillow lava on the shores of Red Indian Lake in Millertown.

**Outcrop Location: N48.81116, W56.54734**

The rocks on the lakeshore at Millertown are made of pillow lava and broken pillow fragments (pillow breccia). Most are subtle shades of greenish grey.

Due to the smoothing action of water over the years, the bulbous surfaces of the pillows are not well preserved. Instead, look for dark ovals in the smoothed rock surfaces – these trace the outline of pillow rims.

The outcrop pictured here is significant because of the mineralization it contains, mainly iron sulphide (pyrite). When the lava erupted on the ocean floor, hot, mineral-rich sea water circulated through the rock, depositing minerals rich in iron, copper, and other elements. From the outcrop you can get a sense for how the fluids moved, seeking out weaker areas of the rock – between pillows or in their still-cooling centres.

Similar ocean-floor mineralization, but on a larger scale, led to significant mineral deposits such as those at nearby Buchans. In fact, much of the mineral wealth of central Newfoundland began as ocean-floor mineralization.

1000          900          800          700          600          500

Z₁               Z₂               Z₃          €

# FYI

- The mineralized rocks in Millertown contain high amounts of copper, but because the deposit is small, it has not been studied in detail.

- The black smokers of today's ocean floor are part of the same process of mineralization by hot fluids.

- The pillow lava at Millertown is part of a curving zone of mineral-rich volcanic rocks known as the Buchans-Robert's Arm volcanic belt (shaded blue in the map below). It extends for more than 200 kilometres from Red Indian Lake to Notre Dame Bay.

## Related Outcrops

On Route 380 beside Crescent Lake just east of Robert's Arm is a pull-off (N49.48704, W55.82733) featuring Cressie, the town's "lake monster" emblem. Across the highway is a large cliff face of pillow lava that is part of the Buchans-Robert's Arm volcanic belt.

See sites 9, 10, and 13 for examples of obvious pillow forms in other parts of Newfoundland.

### Exploring Further

Buchans Miners Museum (N48.82423, W56.85960) on Court Road in Buchans is located in the former mine manager's home and provides information about the rocks, minerals, and mining history of that area.

Pilley's Island Heritage Museum (N49.50586, W55.72480) in Pilley's Island has exhibits and artifacts related to the history of mining in the northern part of the Buchans-Robert's Arm volcanic belt.

"Up in 'Smoke.'" Woods Hole Oceanographic Institute's educational video gallery, www.whoi.edu/VideoGallery. Click on this video to see a black smoker in action.

500    400    300    200    100    0

€ | O | S | D | C | P | Ṯ | J | K | Cz

Southeast Brook flows over Ordovician siltstones and shales north of Milltown-Head of Baie d'Espoir.

# Carpet on the Sea Floor
## Ocean Sediment at Trout Hole Falls

The name suggests this scenic spot could yield a fine, fat fish for supper – and the setting is perfect as rocks gently hold back the water of Southeast Brook to form a quiet pool. But nothing like modern fish existed when the siltstones and shales at Trout Hole Falls formed at the bottom of the Iapetus ocean.

Back then, when jawless, boneless fish roamed the seas, this site was separated from western Newfoundland by the Iapetus ocean. The sediments here were eroded from a landmass near the ancient continent of Gondwana. Western Newfoundland lay in the tropics; Gondwana and the rocks of Trout Hole Falls (along with the rest of Newfoundland east of Red Indian Lake) lay in cooler regions of the southern hemisphere.

The Iapetus ocean basin was complex and tectonically active then – it was in the process of closing. As sediment accumulating on the continental slope near Gondwana became unstable, periodic cascades of silt and clay clouded the water before settling here.

So settle yourself – explore the results of this Ordovician scenario in the rocks at Trout Hole Falls.

# Getting There

### Driving Directions

Follow Route 361 near Milltown-Head of Baie d'Espoir. About 1.5 kilometres northeast of where the highway bridge crosses Southeast Brook, watch for a sign at the entrance to Trout Hole Falls Municipal Park. Turn onto the unpaved park road and follow it for about 200 metres.

### Where to Park

Parking Location: N47.93285, W55.72120

This is a large open gravel area near the entrance to the park.

### Walking Directions

From the parking area, follow the trail and boardwalks northeast toward the brook. Leave the boardwalk near its end point and follow a footpath to the riverbank. Then walk upstream a short distance to the outcrop location.

### Notes

Well-exposed outcrops of shale occur beside the stream bank 10 to 20 metres downstream from the prominent siltstones. They are accessible in low water conditions.

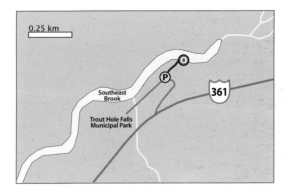

**1:50,000 Map**

St. Alban's 001M13

**Provincial Scenic Route**

Coast of Bays

# On the Outcrop

Resistant siltstone and easily worn shale form alternating layers along Southeast Brook in Trout Hole Falls Municipal Park, Milltown-Head of Baie d'Espoir.

It's easy to recognize the alternating layers of siltstone and shale at this outcrop. They are tilted on edge and lie across the stream like a series of low walls. The shale, being more readily eroded, lies in slots between the prominent beds of siltstone.

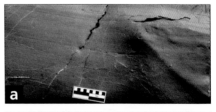

These beds are thought to be a variety of turbidite. The water that deposited them was not swirling as strongly or moving as quickly as for some turbidites (compare with site 6). For that reason, the water here carried no coarse sand, only finer-grained silt and mud. The sediment was deposited in thin, parallel layers with few signs of disturbance (photo a).

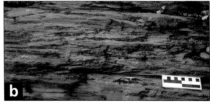

(a) Siltstone (detail); (b) shale (detail).

Each pair of siltstone and shale layers represents a single turbidite event. First the silt settled out, then over time the water cleared as the finer mud particles slowly drifted down.

Later tectonic events tilted and folded the layers, imparting a slight foliation in the process. This effect is best seen in the shales (photo b).

1000　　　　　900　　　　　800　　　　　700　　　　　600　　　　　500

$Z_1$　　　　　　　　　$Z_2$　　　　　　　$Z_3$　　　€

# FYI

- The siltstone and shale along Southeast Brook are part of a thick sequence of ocean sediments. Researchers estimate the pile is about 5 kilometres thick.

- The thick pile of sediment includes some volcanic layers. One volcanic layer, the Twillick Brook formation, is quarried north of Milltown along Route 360. Known as "blue sandstone," it is used in landscaping.

### Related Outcrops

Sediment deposited on the floor of the Iapetus ocean during the Cambrian and Ordovician periods is shaded blue in the map at left.

Oceanview Park (N49.50917, W55.44716) in Leading Tickles offers another look at Ordovician sediments from the Iapetus ocean floor. The southern end of the park is underlain by black shale, while in the northern end are outcrops of turbidite. Both are probably younger than the rocks at Trout Hole Falls. But they all formed on the Gondwanan side of the Iapetus ocean as it was closing during the latter half of the Ordovician period.

Outcrops of sedimentary rock from the Iapetus ocean floor frame the beach at Oceanview Park in Leading Tickles, Notre Dame Bay.

500    400    300    200    100    0

Є    O    S    D    C    P    Ŧ    J    K    Cz

Soapstone outcrops on the hill overlooking the Dorset soapstone quarry and Fleur de Lys harbour.

# Line of Demarcation
## Altered Mantle Rocks at Fleur de Lys

Travellers along the Baie Verte Highway have a silent companion. Snaking in and out, back and forth along Route 410 is a slivered remnant of Iapetus ocean crust and mantle (see Related Outcrops).

The highway follows a major tectonic boundary – the Baie Verte-Brompton Line – separating the Humber and Dunnage zones. The line is marked by ophiolite that was intensely deformed and altered as Iapetus ocean crust was pushed onto the Humber zone (see sites 9-11).

In the Humber zone, mantle rock travelled high atop a thrust stack and is mostly well preserved. But the examples along the Baie Verte-Brompton Line are from a deeper level in the collision process. Heat, pressure, and metamorphic fluids played an important role. As a consequence, most of the mantle rocks along Route 410 were converted to serpentinite, asbestos, soapstone, listvenite, or related rock types.

The 1,500-year-old Dorset soapstone quarry at Fleur de Lys, although slightly west of the main tectonic boundary, provides a good example of the altered mantle rocks so important to the region's history.

# Getting There

### Driving Directions

Follow Route 410 to the east end of Fleur de Lys. Watch for signs for the Dorset soapstone quarry and pull in at the Dorset Museum on the north side of the road (about 200 metres from the end of Route 410).

### Where to Park

Parking Location: N50.11898, W56.12563

Park in the parking lot in front of the Dorset Museum.

### Walking Directions

Walk around the left side of the museum and follow the boardwalk leading behind the building. The outcrop location is the viewing platform by the quarry.

### Notes

This site is under the protection of the Historic Resources Act. Sampling or defacing the outcrops is prohibited. There is no cost to visit the archeological site, but the museum charges a fee to view the exhibits.

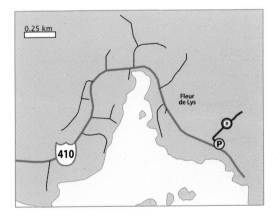

---

**1:50,000 Map**

Fleur de Lys 012I01

**Provincial Scenic Route**

Dorset Trail

# On the Outcrop

There are numerous outcrops of soapstone around the quarry in Fleur de Lys: (a) The complex web of veins visible on an unquarried rock face channelled fluids during formation of the soapstone; (b) an outcrop near the museum provides a closer look at the soapstone's "elephant skin" surface texture; (c) extraction scars in quarried rock preserve evidence of stone age craft.

## Outcrop Location: N50.11992, W56.12507

The Dorset Museum in Fleur de Lys is surrounded to the east and north by large outcrops of soapstone. Heat, deformation, and exposure to hot, mineral-rich fluids changed the chemical and mineral composition of the rock, creating soapstone from peridotite.

Soapstone is mostly made of talc, a magnesium-rich mineral. Talc gives soapstone many of its useful properties, for example, its softness, heat capacity, and resistance to chemicals.

The conversion of peridotite to soapstone occurs along cracks where fluids invade the rock. The formation of soapstone increases the rock volume and, because talc is weak, allows the rock to deform more easily. Expansion and deformation create new cracks; new cracks let more fluids permeate the rock; and so the process continues. The texture of the natural, uncarved rock surfaces at this site resulted in part from the fracturing and expansion of the rock on many different scales.

1000    900    800    700    600    500
Z₁         Z₂         Z₃      Є

# FYI

- The Advocate asbestos deposit near Baie Verte and the Thetford asbestos deposit in Quebec both lie along the Baie Verte-Brompton Line.

- The chemistry of mantle rocks in the Baie Verte region suggests they are all part of a single slab that formed about 490 million years ago. Alteration occurred when the rocks were pushed onto Laurentia about 480 million years ago.

The Advocate asbestos mine north of Baie Verte was an open pit operation.

### Related Outcrops

On the Baie Verte peninsula, ophiolitic rocks (shaded blue in the map at right) including fragments of mantle occur in deformed slivers along the Baie Verte-Brompton Line and in less deformed areas to the east.

On Route 410 about 9 kilometres north of the intersection with Route 412, there is a pull-off on the east side of the highway (N49.97990, W56.19828) beside the old Advocate asbestos mine. The site affords a good view of the open pit, when weather conditions permit.

### Exploring Further

Baie Verte Peninsula Miner's Museum (N49.92137, W56.22971) is located near the intersection of Routes 410 and 412 in Baie Verte. The museum contains large mineral specimens and exhibits related to the history of mining on the peninsula.

Steeply dipping greenstone buttresses the shore beside the Petit Nord picnic site in Coachman's Cove, Baie Verte peninsula.

# End of the Line

## Ophiolitic Melange at Coachman's Cove

Coachman's Cove is steeped in history as part of the "Petit Nord" of Newfoundland's French Shore. To commemorate the French connection, a traditional bread oven has been created in the little park at this site.

Not only historians but also geologists know this cove well. Over the past several decades the "picnic site locality," as it is known to geology students, has found its way into many field guides and become a classic field-trip stop.

No wonder. It includes an unusual rock type, showcases complex structures, and is the end point in North America of the Baie Verte-Brompton Line – a tectonic boundary separating the Humber and Dunnage zones of the Appalachian mountain belt (also see site 18).

Like the melanges of the Humber zone (see sites 6-9), this one at Coachman's Cove formed when ocean crust was pushed onto the margin of the continent Laurentia. In this case, the melange itself is ophiolitic – it is made of rocks derived from the ocean floor.

# Getting There

### Driving Directions

Follow Route 410 to the area about 17 kilometres north of Baie Verte and watch for the intersection with Route 410-10 to Coachman's Cove. Follow Route 410-10 into Coachman's Cove and around the south side of the cove. Continue as the road changes to gravel, travelling over a short causeway onto the headland and into the parking area.

### Where to Park

Parking Location: N50.05035, W56.10715

Park in the gravel parking lot for the picnic area.

### Walking Directions

From the parking area, walk down the steps and across the grass, past the historic oven site to the shore. Walk toward the right along the shore to the outcrop location. To explore the site further, walk back through the parking area and follow the gravel track over to the north side of the peninsula.

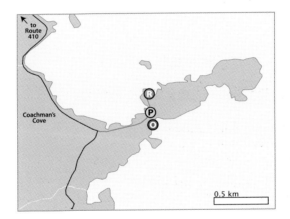

---

**1:50,000 Map**

Fleur de Lys 012I01

**Provincial Scenic Route**

Dorset Trail

# On the Outcrop

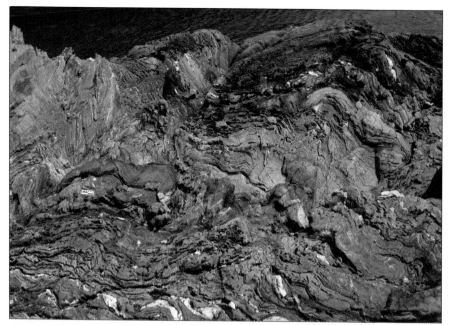

Complex fold patterns and a mixture of ocean-floor rock types create eye-catching outcrops in parts of the Coachman's Cove melange.

## Outcrop Location: N50.04987, W56.10697

At Coachman's Cove you'll find a variety of metamorphic rock types.

Greenstone is common and can be recognized by its grey green colour. It formed from ocean-floor volcanic eruptions, perhaps originally pillow lava. Much of the greenstone here stands in steeply tilted arrays with strong layering due to foliation.

Among the greenstone are multicoloured rock layers with complicated fold patterns. Researchers originally attributed all of this folding to the obduction of Iapetus crust onto Laurentia. Now much of the deformation is known to have occurred during a later, Silurian orogeny (see sites 20-24).

A distinctive rock type at this site is coticule. Coticule is rich in iron and manganese and contains a lot of garnet, which gives the rock a reddish colour. Though its origin is uncertain, it may have begun as a metal-rich chert on the Iapetus ocean floor, probably near a mineral-rich hydrothermal vent.

Other rock types found at this site include serpentinite, marble, and a black, graphite-bearing schist – all originally part of the Iapteus ocean floor.

1000    900    800    700    600    500

$Z_1$    $Z_2$    $Z_3$    $\in$

# FYI

- The Baie Verte-Brompton Line continues in Europe as the Highland Boundary Fault of Scotland.

- Ophiolitic rock in the eastern parts of the Baie Verte peninsula is not strongly deformed. During obduction, intense deformation was restricted to the Coachman's Cove melange, just as it was in the melanges between layers in the Humber Arm allochthon (see sites 6-9).

- Coticule is found throughout the Appalachians, usually in rock formations of Cambrian and Ordovician age. It is also found in the Caledonian mountain belt, Europe's equivalent of the Appalachians. Belgian varieties of this rock are especially prized as whetstones for precision sharpening tasks. In fact, *coticula*, in Latin, means whetstone.

## Related Outcrops

Farther east, the boundary between the Dunnage and Gander zones is called the Gander River Ultramafic Belt (GRUB). Like the BaieVerte-Brompton Line (BV-BL), it is a zone of highly deformed, steeply inclined rocks, with altered slivers of mantle rock distributed along its length (see site 27). Both linear zones of mantle rock (shaded blue in the map below) formed as ocean crust was pushed onto a nearby continent – westward onto Laurentia along the Baie Verte-Brompton Line and eastward onto Gondwana along the Gander River Ultramafic Belt.

Another of the Dunnage zone's well-known melanges (the Dunnage melange, see site 20) occurs on and near New World Island. It relates to a later phase in the closure of the Iapetus ocean.

500    400    300    200    100    0

€  O  S  D  C  P  Ŧ  J  K  Cz

Summerford Hiking Trail on New World Island lies entirely within the Dunnage melange. Some of the hummocks and hills along the trail – like the one just past the bridge shown here – are made of resistant blocks within the weaker shale.

# Blocks of All Sizes

## Dunnage Melange at Summerford Trail and Lookout

The Summerford Hiking Trail lies within the Dunnage melange. Its peaceful, well-worn paths belie the origin of the rocks underfoot: This area was the scene of a titanic drama about 470 million years ago.

The mid-ocean ridge that produced the Iapetus ocean floor was sinking into a subduction zone. In a long, narrow basin nearby, fine sediment accumulated to form deep layers of submarine mud.

A wide assortment of rocks tumbled into the basin from adjacent areas as earthquakes shook the terrain. Gravel and boulders, rocks the size of buildings, and even fragments as large as whole city blocks, fell into the soft mud. As the mid-ocean ridge descended, magma rose into the basin and mixed with the soft mud as well.

The mud of the basin became the dark shale of the Dunnage melange, visible along this trail. As for the blocks – you'll climb one to the lookout. From there, you can see the remains of many other large blocks in a breathtaking view of Dildo Run.

# Getting There

### Driving Directions

Follow Route 340 to Summerford on New World Island. About 0.25 kilometres south of the intersection with Route 344, watch for a sign for the Summerford Hiking Trail and turn northeast into the parking area.

### Where to Park

Parking Location: N49.49784, W54.77715

This is a small gravel parking area at the trail head for the Summerford Hiking Trail.

### Walking Directions

From the parking area, cross a wooden bridge and walk over a small rise to reach the main trail. The trail encircles a pond; keep left for the most direct route to the lookout. Watch for a wooden stairway on your left and climb to the viewing platform.

### Notes

This is a community hiking trail.

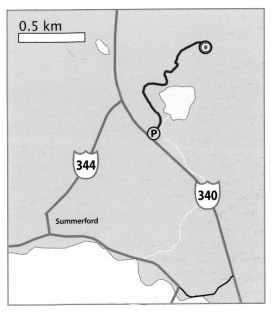

---

**1:50,000 Map**

Comfort Cove-Newstead 002E07

Twillingate 002E10

**Provincial Scenic Route**

Kittiwake Coast: Road to the Isles

# On the Outcrop

Melange blocks and areas of igneous rock form islands in Dildo Run, as seen from the Summerford lookout.

**Outcrop Location: N49.50167, W54.77367**

As you walk along the Summerford Hiking Trail, you'll encounter numerous outcrops of dark, flaky shale like the one pictured below. It is the matrix of the Dunnage melange, the geological gravy of this tectonic stew. The shale is weak and easily eroded – you'll find it in the low ground along the trail.

From the lookout, you can see the distribution of the shale on a larger scale: Much of it is under the waters of Dildo Run, having been deeply eroded by glacial action.

The lookout provides a wide panorama of melange blocks and areas of igneous rock in the form of islands, large and small. The "meat and potatoes" of the Dunnage melange, these

Dark, flaky shale forms low outcrops along the Summerford Hiking Trail.

blocks and intrusions have provided many clues to the setting that brought them together.

From Summerford lookout you can get a feel for the scale of what happened as the melange formed.

1000          900          800          700          600          500

Z₁                          Z₂                    Z₃          €

# FYI

- The Dunnage melange extends about 40 kilometres from New World Island southwestward into the Bay of Exploits. It is one of the most extensive melanges in the Appalachian mountain belt.

- The melange lies just south of the Red Indian Line, a major tectonic boundary. The line extends from Cape Ray in the southwest to Triton Island in the north, then bends eastward from Badger Bay to New World Island. It separates Iapetus ocean rocks that formed near Laurentia (shaded light blue on the map above) from those that formed near Gondwana (shaded dark blue on the map above).

## Related Outcrops

Newfoundland is host to many block-in-matrix melanges like the Dunnage melange. For example, the ones at Rocky Harbour (see site 7) and L'Anse aux Meadows (see site 8) are both sedimentary melanges with a shale matrix.

Bare, rough outcrops of coarse conglomerate (foreground and distance) cap the high ground on both sides of Pike's Arm.

# Uplift and Erosion
## Silurian Conglomerate at Pike's Arm

There are plenty of reasons to get yourself all the way up onto Pike's Arm lookout. They include a great view (perhaps including icebergs), plenty of blueberries (in season), and the amazing variety of rock types in this colourful conglomerate.

Moreover, the rock here is part of a remarkable story. As the Iapetus ocean closed, parts of Laurentia and Gondwana came into contact. Laurentia rode over Gondwana, so Laurentia was uplifted and became mountainous, while a narrow seaway remained on the Gondwanan side where Pike's Arm is situated.

The Red Indian Line separating Laurentia from Gondwana runs along Goshen Arm, right next to the peninsula you are climbing on. Picture it: The sand, grit, cobbles, and boulders in the conglomerate travelled – tumbled, really – from newly formed mountains in Laurentia, down into the narrow remains of Iapetus where you are standing.

The conglomerate here is near the top of a thick sequence of sediments that filled in the seaway over time. At the end of the process, this part of the Iapetus ocean was gone.

# Getting There

## Driving Directions

Follow Route 346 toward its termination on New World Island. About 1.75 kilometres before the bridge in Toogood Arm, turn onto Route 346-10 and follow it into the town of Pike's Arm. Watch for signs to the Pike's Arm lookout. North of the bridge in Pike's Arm, fork left. About 300 metres beyond the fork, a short gravel drive on the north side of the road leads up into the parking area.

## Where to Park

Parking Location: N49.64715, W54.58303

This is a small gravel parking lot at the trail head.

## Walking Directions

Follow the trail and stairways up the hill. There is a viewing platform at the top as well as many good outcrops along the way.

## Notes

This is a community hiking trail.

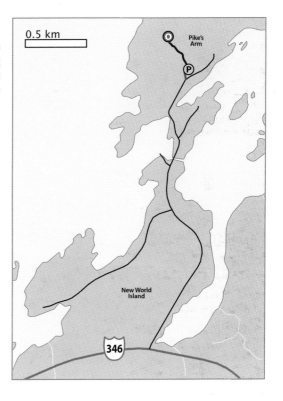

---

**1:50,000 Map**

Twillingate 002E10

**Provincial Scenic Route**

Kittiwake Coast: Road to the Isles

# On the Outcrop

Rock fragments of many colours, sizes, and shapes are bound in a sandy matrix in the conglomerate at Pike's Arm.

Outcrop Location: N49.64870, W54.58455

The outcrop location is at the lookout. As you climb, you will see various forms of the conglomerate. Near the bottom of the stairway, it is gravelly, with occasional large cobbles. Farther up the hill, the average size of the fragments increases.

Make yourself comfortable on an outcrop and take a close look at the individual bits and pieces the rock contains. You'll see the collection is poorly sorted, meaning it contains fragments of many different sizes. Some of the fragments are rather angular, while others are more rounded and smooth. The wide range of sizes and the angular fragments are both characteristic of sediments that were eroded and deposited quickly – in this case, probably by underwater landslides, or debris flows.

The great variety of rock types in the conglomerate at Pike's Arm makes it unique in the region. You'll see greenish volcanic rocks, bright red chert, mottled grey granitic fragments, white quartzite, and others. They are typical of the rocks found on the Laurentian side of the Red Indian Line.

| 1000 | 900 | 800 | 700 | 600 | 500 |
|------|-----|-----|-----|-----|-----|
| $Z_1$ | | $Z_2$ | | $Z_3$ | $\varepsilon$ |

# FYI

- The conglomerate at Pike's Arm is about 600 metres thick. Its features suggest it was deposited in an underwater delta at the base of a slope.

- Zircon mineral grains in the sandy matrix of similar conglomerates were studied to find out where the sediment came from. Most of the zircons had ages typical of Iapetus ocean rocks (470-490 million years). A few older zircons came from Grenville-age rocks like those in the Long Range mountains (see site 1).

- From lower to higher layers in the conglomerate, there is a gradual increase in the amount of granitic fragments. That suggests volcanic arcs were being unroofed, that is, eroded more and more deeply until the granitic intrusions beneath them were exposed.

## Related Outcrops

Late Ordovician to Early Silurian melange, conglomerate, and turbidite (shaded blue in the map below) are found south of the Red Indian Line all along Notre Dame Bay, from Sop's Head in the west to Dog Bay in the east. They appear to be related to uplift and erosion of the Laurentian terrain to the north as Iapetus closed along the Red Indian Line.

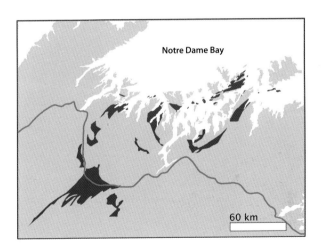

500    400    300    200    100    0

€  O  S  D  C  P  T̶  J  K  Cz

Red pebbles of volcanic lava and ash line the shore at Goodyear's Cove, Hall's Bay.

# Fiery Cauldron

## Springdale Caldera at Goodyear's Cove

The Iapetus ocean threw itself a going-away party during the Silurian period and went out with a bang. As the last seaway disappeared along the Dog Bay Line farther east (see site 23), continental collision wrenched the crust of the Dunnage zone. An already subducted slab of Iapetus ocean crust broke off in the mantle, exposing the base of the colliding continents to immense heat from below.

In west-central Newfoundland this led to the formation of a group of volcanoes that showered the region with lava and ash. The volcanoes cratered the landscape with a series of explosive eruptions forming collapse structures called calderas.

Calderas form when the magma chamber below a volcano is emptied suddenly, usually by a huge outpouring of volcanic ash. The weight of the overlying volcano is no longer supported, and it crashes down into the empty chamber. The best-known caldera in North America is probably the one at Yellowstone National Park in Wyoming (US).

The Springdale caldera is estimated to have been about 65 kilometres long by 35 kilometres wide. At Goodyear's Cove you can camp, hike, or picnic on part of it and take a close look at rocks from two kinds of eruptions.

# Getting There

### Driving Directions

Follow the Trans-Canada Highway to the area around Hall's Bay. About 1.25 kilometres east of the intersection with Route 380, watch for the entrance to Goodyear's Cove park and campground. Turn onto the park's gravel road on the north side of the highway and stop at the chalet to register. After registering, or if the chalet is unattended, proceed downhill into the park.

### Where to Park

Parking Location: N49.42769, W56.10414

Park in the gravel area along the shore, or as indicated by the park attendant.

### Walking Directions

As you face the cobble beach, there is a wooden stairway into the woods on the left. This is the trail head for the Goodyear's Cove Hiking Trail. Proceed along the trail, boardwalks, and stairways, following signs for a "bird's eye view of South Brook." The outcrop is at the viewing platform.

### Notes

Goodyear's Cove is a campground and park maintained by the community of South Brook. There are fees for camping and day use during the camping season. If the chalet is unattended when you arrive, but appears to be open, please check back on your way out.

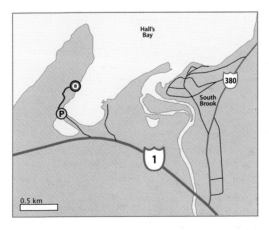

---

**1:50,000 Map**

Springdale 012H08

**Provincial Scenic Route**

Green Bay and the Beothuk Trail

# On the Outcrop

Rock fragments caught up in a volcanic explosion are embedded in reddish volcanic ash in this boulder on the shore of Goodyear's Cove. The same rock formation has a similar but more weathered appearance on the outcrop at the lookout.

## Outcrop Location: N49.42966, W56.10266

To see the volcanic breccia in its original position, climb to the outcrop location at the lookout. The rock surfaces there are quite weathered, but you can see the breccia's texture clearly. Then, if the tide conditions allow, walk along the shore around the west side of the cove. There, some large boulders have fallen from above. Their broken surfaces are fresher, so the colours are more vivid and detailed.

The breccia consists of angular fragments in a fine-grained matrix of volcanic ash. The fragments were broken away from the walls of the volcanic vent and picked up by the hot, explosive eruption as it rushed out past existing volcanic deposits. The overall rock composition is rhyolite.

Boulders of a more even-textured andesite lava appear on the beach. Many examples contain visible white feldspar crystals in a brick red matrix. These come from the layer below the breccia. Other evidence from the area around Goodyear's Cove suggests that the lava and breccia formed part of a volcanic dome built of lava and ash flows.

1000   900   800   700   600   500

$Z_1$   $Z_2$   $Z_3$   $\in$

# FYI

- The dimensions of the Springdale caldera suggest that it may have erupted between 1,000 and 10,000 cubic kilometres of ash.

- The Iapetus ocean was gone by this time; all the volcanic materials from the caldera were deposited on dry land.

## Related Outcrops

Two fine hiking trails near King's Point traverse areas of volcanic ash closely related to the rocks at Goodyear's Cove.

- Alexander Murray Hiking Trail (trail head N49.57948, W56.19572), 9 kilometres. Volcanic ash deposits are in the higher, inland sections of the trail.

- Rattling Brook Hiking Trail (trail head N49.62084, W56.17606), 0.5 kilometres. The brook and waterfall both flow across volcanic ash deposits.

The 800-metre-high waterfall at Rattling Brook, near King's Point, flows across Silurian volcanic ash formations.

Silurian volcanic rocks and accompanying intrusions (shaded blue in the map at left) are widespread in west-central Newfoundland.

West of Tilting harbour, Turpin's Trail crosses dark mafic and ultramafic rocks of the Tilting layered suite.

# Beneath a Volcano
## Relict Magma Chamber at Tilting, Fogo Island

The town of Tilting on Fogo Island is significant in many ways. It was the first Provincial Heritage District created in Newfoundland. As well, Parks Canada has designated Tilting as a National Cultural Landscape District. The town's unique cultural heritage includes many long-established walking trails used by local residents and visitors from around the world.

The trails wind over wonderful exposures of the Fogo Island batholith, a complex intrusion that underlies 80 per cent of the island. Like several other Silurian bimodal intrusions in Newfoundland, the Fogo Island batholith contains both dark-coloured mafic and lighter-coloured granitic portions. But it also contains ultramafic rocks, which is unusual. All three rock types can be seen in Tilting.

The rocks formed as complex movement between continental masses wrenched the crust of the Dunnage zone. The Iapetus ocean was gone, its last, narrow seaway just closing along Dog Bay Line (see Related Outcrops). Mafic and granitic magmas pooled and gently circulated in a stable chamber just below the Earth's surface. Unlike the explosive calderas in west-central Newfoundland (see site 22), the Fogo Island batholith preserves a slow, quiet process. Peaceful Tilting is a perfect place to observe the outcome.

# Getting There

### Driving Directions

On Fogo Island, follow Route 334 to the town of Tilting. Two parking locations provide access to Turpin's Trail; a third is in Oliver's Cove.

1a. Watch for signs to Bunker Hill. Turn north onto the gravel road and follow it for about 0.25 kilometres to the parking area.

1b. Turn onto Kelly's Point Road in Tilting and follow it along the west side of the harbour.

2. By the south end of Tilting harbour, turn south onto Poore's Lane and follow it about 0.4 kilometres to Oliver's Cove.

### Where to Park

Parking Locations:

Bunker Hill: N49.70876, W54.07176

Kelly's Point Road: N49.70763, W54.06430

Oliver's Cove: N49.69917, W54.05534

### Walking Directions

For Turpin's Trail, the eastern trail head near Kelly's Point Road is next to the Lane House Museum. Start from either parking location and follow the trail between the lighthouse and Bunker Hill to see a wide variety of outcrops.

In Oliver's Cove, follow the footpath west along the shore for about 150 metres to the outcrop.

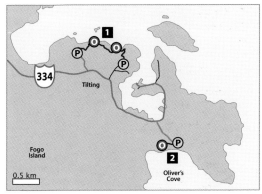

---

**1:50,000 Map**

Fogo 002E09

**Provincial Scenic Route**

Kittiwake Coast: Islands Experience

# On the Outcrop (1)

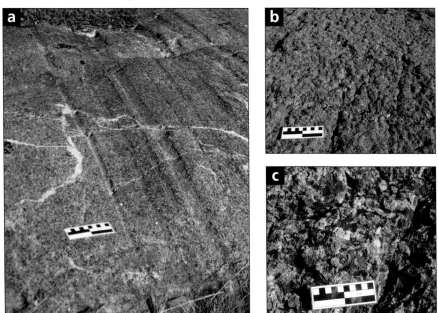

Different rock types of the Tilting layered suite can be seen along Turpin's Trail in Tilting, Fogo Island: (a) layered gabbro; (b) a lumpy ultramafic rock, websterite; (c) gabbro of large hornblende and feldspar crystals.

**Outcrop Location: N49.70907, W54.06488**

Between the lighthouse west of Tilting harbour and Bunker Hill, Turpin's Trail crosses rocks characteristic of the Tilting layered suite, one of the mafic zones of the Fogo Island batholith. The layered suite formed from magma in which crystals settled into layers, much like sand grains settle in still water.

At the outcrop location, about 100 metres west of the lighthouse, are good examples of two significant rock types. One, a layered gabbro (photo a), is the most common rock in the layered suite. Nearby is a distinctive ultramafic rock called websterite (photo b). It formed when dark pyroxene crystals sank and accumulated in a low-lying layer in the magma chamber.

About 150 metres east of Bunker Hill (N49.70988, W54.06967), another dramatic rock type will catch your eye (photo c). Here, the gabbro is made of large dark and light crystals that formed near the top of the sequence.

| 1000 | 900 | 800 | 700 | 600 | 500 |
|---|---|---|---|---|---|
| $Z_1$ | | $Z_2$ | | $Z_3$ | € |

# FYI

- The Fogo Island batholith was a slab-like intrusion between 6.5 and 8 kilometres thick. It was intruded into a shallow level of the crust. Its relationship to the volcanic rocks of Brimstone Head in the town of Fogo is uncertain.

- There are several cycles of websterite and gabbro in the Tilting layered suite. That suggests several pulses of new magma entered the chamber without disturbing previous layers, then went through the same process of crystals forming, sinking, and accumulating.

## Related Outcrops

Four other igneous intrusions (Mount Peyton, Hodges Hill, Long Island, and Loon Bay) are of similar age and occur roughly in line with the Fogo Island batholith. All five (shaded blue in the map at right) are Silurian-Devonian bimodal intrusions between the Dog Bay Line and Red Indian Line, so they may have a common origin. However, their rock chemistry suggests they did not all form in the exact same way.

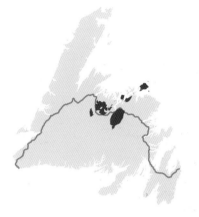

Mafic and granitic dykes along the Trans-Canada Highway northwest of Badger (N49.09880, W56.07554) are part of a bimodal Silurian dyke swarm that occurs over a wide area in central Newfoundland.

500    400    300    200    100    0

Є    O    S    D    C    P    Ṟ    J    K    Cz

# On the Outcrop (2)

In Oliver's Cove south of Tilting, the rocks are lighter in colour and belong to the granitic phase of the Fogo Island batholith.

## Outcrop Location: N49.69900, W54.05770

To get to Oliver's Cove you pass through an area of Tilting that, through long tradition, has been fenced and developed into pastures and family gardens. Please stay on the established walking trails.

The rocks in this part of the town have a very different appearance from those along Turpin's Trail. The best exposures can be seen at low tide along the shore, where the rock has an evenly mottled grey appearance. Several granitic rock types surround the Tilting layered suite. This one is not a true granite, but rather a tonalite.

Oliver's Cove tonalite (detail).

Trails lead along both sides of Oliver's Cove. If you explore them, you'll find the whole area is underlain by similar granitic rock. Occasionally you may see fragments of other rock types within the intrusion, or areas where the mineral grains are more diverse in size. But the overall composition of the rock varies little from the example at the outcrop location noted above.

| 1000 | 900 | 800 | 700 | 600 | 500 |
|---|---|---|---|---|---|
| $Z_1$ | | $Z_2$ | | $Z_3$ | € |

# FYI

- Some igneous intrusions show signs of abundant fluids, for example, mineral veins, areas of very large crystals, or signs of violent eruptions. In contrast, the Fogo Island batholith appears to have formed from a relatively dry magma.

- The batholith was slightly folded and tilted on its side by later events, which is why so many of its features are exposed today.

- Like the calderas and related intrusions of west-central Newfoundland (see site 22), the Fogo Island batholith may provide evidence that a complex, prolonged continental collision caused melting deep in the crust of the Dunnage zone during the Silurian and Devonian periods.

## Related Outcrops

While travelling to or from the Change Islands-Fogo ferry terminal on Route 355, you can stop in Stoneville for views across Dog Bay. For example, the view from the Pentecostal Church parking lot (N49.45570, W54.54590) is very scenic.

Dog Bay is significant because a tectonic boundary, the Dog Bay Line, runs along it. Rocks on either side of the line have different histories. They were separated until late in the Silurian period by the last remaining basin of the Iapetus ocean. Once subduction closed that basin, Iapetus was gone.

A significant tectonic boundary lies under the waters of Dog Bay, seen here from the shore in Stoneville.

Reddish river sandstone outcrops along the shores of the Exploits river in the town of Bishop's Falls.

# Ocean No More
## River Sandstone at Bishop's Falls

For almost 120 million years, the Dunnage zone was an ocean realm, a world of pillow lava, sea-floor sediment, and isolated ocean islands. From late in the Ediacaran period until the middle of the Silurian period, creation and destruction of ocean crust was the be-all and end-all of this region.

The sandstone at Bishop's Falls marks a fundamental change in the landscape. By the time it formed, Iapetus had closed. Where sea had lapped shore along the Red Indian Line and Dog Bay Line, now land met land. It was a freshwater world.

The signal rock type for this transformation is a brick red river sandstone that appears extensively in east-central Newfoundland. For example, along the Trans-Canada Highway between Grand Falls and Bishop's Falls, or along Route 360 between Bishop's Falls and Botwood, you'll find the characteristic colour in outcrops along the highway and sometimes in the road surface itself.

What could be more appropriate than to examine these river sandstones along the banks of the beautiful Exploits? Fallsview Municipal Park in Bishop's Falls awaits.

# Getting There

## Driving Directions

From the Trans-Canada Highway east of Grand Falls, take exit 21 or 22 and follow Main Street in Bishop's Falls. Watch for signs to Fallsview Municipal Park. About 2 kilometres west of exit 22, turn south onto Powerhouse Road and proceed to the park entrance. Stop at the kiosk to register, then drive down the hill into the park.

## Where to Park

Parking Location: N49.01633, W55.47064

Park on the lower level of the park, by the river.

## Walking Directions

The outcrop is beside the lower level parking area, along the river bank at the upstream end of the park.

## Notes

This site is located within the Fallsview Municipal Park. During the camping season, there are fees for day use and camping.

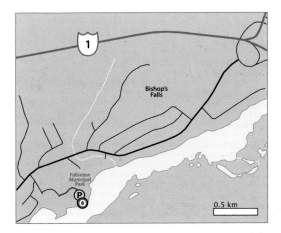

---

**1:50,000 Map**

Botwood 002E03

**Provincial Scenic Route**

Exploits Valley

# On the Outcrop

Colour variations highlight bedding and cross-bedding in sandstone at Fallsview Municipal Park, Bishop's Falls.

## Outcrop Location: N49.01599, W55.47063

In Fallsview Municipal Park, the beds of river sandstone are tilted on edge. The outcrops have been sculpted and smoothed over time by a combination of glacial action and erosion by Exploits' currents, highlighting the beds' colour variations.

Colour banding in the sediments also allows you to see details geologists use to figure out the environment in which sediments were deposited.

In places on the outcrop, you can see trough-like patterns in which the fine layering sweeps into a truncated curve. This is called cross-bedding.

River sandstone (detail).

It forms as ripples, or miniature dunes in the sediments, are moved around by flowing water. You can also see the alternation between layers of fine silt and coarser sand. This is caused by water flowing at different rates, perhaps as the river channel shifted back and forth over time.

1000   900   800   700   600   500

Z₁   Z₂   Z₃   €

# FYI

- The red of the sandstones is typical of river-deposited sediments or other shallow-water environments where plenty of oxygen is present. That allows iron in the sediments to be oxidized – like rust – giving it a reddish hue.

- The sandstones around Grand Falls, Bishop's Falls, and Botwood have sedimentary structures that include raindrop imprints and mud cracks, clearly indicating that the sediments were deposited on land rather than in the sea.

- Zircon mineral grains in sediment preserve information about the age of the rocks the sediments came from. Many zircons in the river sandstones of Grand Falls and Botwood have ages ranging from 1,500 to 1,000 million years, typical of pre-Iapetus Laurentia (see site 1).

## Related Outcrops

Silurian river sediments (shaded blue in the map below) are widespread in the Dunnage zone. They are especially abundant in a wide zone that includes Grand Falls, Bishop's Falls, and Botwood. In fact, the sandstones of this age in Newfoundland are referred to as the Botwood group.

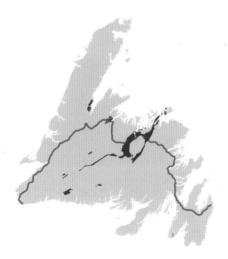

At the Botwood Heritage Centre (N49.15197, W55.34110) there is a causeway out to Killick island in the harbour. There you can stroll on more examples of the sandstone formed by the erosion of Laurentia by Silurian rivers.

## Exploring Further

US Geological Survey bedform sedimentology site, walrus.wr.usgs.gov/seds/bedforms. The site has animations showing a variety of ways in which cross-bedding forms.

500    400    300    200    100    0

€  O  S  D  C  P  Ŧ  J  K  Cz

Cliffs of Ordovician sediment host a small number of Jurassic dykes along Leading Tickles, seen here from the south shore.

# Opening Again
## Jurassic Dykes at Leading Tickles

Between 1965 and 1968 a revolution in thinking swept through the geological community. Where previously the Earth had been considered static, within a few years' time the concept of a mobile, constantly changing arrangement of tectonic plates was widely accepted.

Canadian geophysicist J. Tuzo Wilson in many ways led this revolution by drawing together diverse lines of evidence into a coherent theory. He outlined how "a succession of ocean basins may have been born, grown, diminished, and closed again" as the means by which mountain belts form. This idea became central to the theory of plate tectonics and is now known as the Wilson Cycle.

In Leading Tickles, you can pay homage to Professor Wilson by visiting a remarkable outcrop that succinctly illustrates the Wilson Cycle in action. At this site, sedimentary rocks from the Iapetus ocean – now part of the Appalachian mountain belt – are cut by lamprophyre dykes intruded as the modern-day Atlantic ocean began to open millions of years later.

# Getting There

### Driving Directions

Follow Route 350 to the area around Leading Tickles, south of the bridge to Cull Island. About 0.5 kilometres west of the bridge, watch for a site south of the highway where the rock cliff has been cut back to form a pull-off.

### Where to Park

Parking Location: N49.50077, W55.44903

This is a pull-off beside Route 350.

### Walking Directions

The outcrop is in the rock wall beside the pull-off.

### Notes

A second outcrop occurs in a pull-off about 100 metres east of the parking location. Walk or drive east to view it.

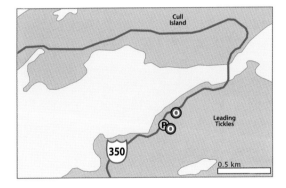

---

**1:50,000 Map**

Point Leamington 002E06

Exploits 002E11

**Provincial Scenic Route**

Exploits Valley

# On the Outcrop

Weathered brown, lamprophyre dykes cut across Ordovician sediments along the harbour in Leading Tickles: (a) at the main outcrop location; (b) nearby, about 100 metres to the east.

**Outcrop Location: N49.50077, W55.44903**

The main outcrop is located in a sort of alcove cut into the cliff face along the highway. The dyke cuts vertically through the surrounding rock. It has weathered to a velvety brown, in striking contrast to the dark red rocks on either side.

The sides of the dyke are roughly parallel, with sharp, clean edges. If you look closely along the edge of the dyke, you can see a narrow zone where the minerals are very fine-grained compared to those in the centre of the dyke. This is a chilled margin, formed when molten rock cooled quickly (with little time for crystals to grow) against the rocks on either side.

The nearby dyke is also nearly vertical and weathered brown.

1000     900     800     700     600     500

$Z_1$     $Z_2$     $Z_3$     $\mathcal{C}$

# FYI

- The dykes at this site are part of the youngest known igneous event in Newfoundland. At the time they formed, the supercontinent Pangaea was beginning to break up.

- Dyke swarms and related igneous intrusions formed all along the fracture zone where the Atlantic ocean opened. The intrusions are now found in North and South America, Europe, and Africa.

- In North America, lamprophyre dyke swarms similar to those at Leading Tickles occur in Quebec and New England.

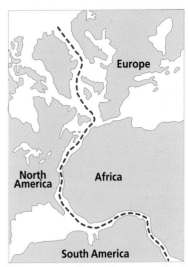

The supercontinent Pangaea rifted apart when the Atlantic ocean opened during the Jurassic period.

## Related Outcrops

The dykes in Leading Tickles are part of a swarm of more than 100 dykes that may be associated with a nearby Jurassic intrusion, the Budgell Harbour stock. Other similar dykes are distributed across the Notre Dame Bay region.

The dykes at Western Brook Pond (see site 1) and the ones at this site are separated by hundreds of millions of years in time, but they formed in similar settings. In each case, the dykes were intruded as a supercontinent (first Rodinia, then Pangaea) rifted apart to form a new ocean (first Iapetus, then the Atlantic).

## Exploring Further

US Geological Survey website, www.usgs.gov. To watch an animation of the Atlantic ocean opening, visit the site and search for "Pangaea breaks up."

# GANDER
## *Zone at a Glance*

### Boundaries

West: GRUB Line
East: Dover-Hermitage Bay Fault

### Origin

Continental margin of Gondwana

### Characteristic Features

Iapetus ocean: sediments–crust–mantle
Continental margin: sandstone–siltstone
Metamorphism: schist and gneiss
Closure–collision: granites

## In the Gander zone, you can …

| 26 | Gander | Visit rocks that formed on a far-away continent. |
|---|---|---|
| 27 | Little Harbour | Learn to recognize an unusual rock type from the mantle. |
| 28 | Margaree | Unscramble the origins of a complex gneiss. |
| 29 | Port aux Basques | Explore a colourful schist formed deep in the Earth. |
| 30 | Rose Blanche | Get acquainted with granite at Newfoundland's only stone lighthouse. |
| 31 | Windmill Bight | Delve into a shear zone, where stressed rock flowed instead of breaking. |
| 32 | Greenspond | Trek through a sea of granite formed during continental collision. |
| 33 | Dover | Sight along the trace of a continental collision. |
| 34 | Lumsden | Contemplate the evidence of tectonic peace and quiet. |

The Gander zone originated somewhere along the continental margin of Gondwana. Between about 540 and 480 million years ago, sand, silt, and mud piled up on a continental slope as the Iapetus ocean grew wider. At that time the Gander zone lay in cool regions of the southern hemisphere.

Later tectonic events separated the Gander zone from the rest of Gondwana, so it's not clear exactly where the sediment came from. West Africa or Amazon regions of South America are possible sources. The sediment is best preserved today in the area around Gander (site 26), where it is only slightly metamorphosed.

Port aux Basques (see site 29).

About 480 million years ago, early stages of ocean closure caused a wide swath of Iapetus ocean crust and mantle to be pushed on top of the Gander zone from the west (site 27). In fact, much of the eastern Dunnage zone is a sort of layer cake with Dunnage rocks above and Gander rocks below.

As the Iapetus ocean continued to evolve, rifting, subduction, and accompanying volcanic activity created a "tectonic collage" in areas where the Gander and Dunnage zones intermingled (site 28).

Rose Blanche (see site 30).

Greenspond (see site 32).

One of the things that makes the Gander zone interesting to explore is that some areas now exposed at the surface were once deeply buried in the Earth's crust. (Much of the rest of Newfoundland has never been deeply buried.) This happened during final closure of the Iapetus ocean about 420 million years ago. In the southwest, ordinary sand, silt, and mud have been transformed by metamorphism into colourful schists containing red garnet, blue kyanite, and large, shiny plates of black biotite (site 29).

After the Iapetus ocean closed, continental collision wrenched the crust, causing large volumes of syn-tectonic granite to form (sites 30 and 32). The granite is partly formed of Gander zone sediment that was so severely metamorphosed that it melted.

The Gander zone was caught in a complex tectonic scrum between 420 and 400 million years ago. The newly arrived microcontinent, Avalonia, wrenched and ground against the eastern margin of the Gander zone. First deep crustal shearing occurred (site 31) followed by shallower faulting (site 33) as the Gander and Avalon zones were joined.

Deadman's Bay (see site 34).

By 385 million years ago, motion between the Gander and Avalon zones had ceased. Melting deep in the newly welded, stable crust formed post-tectonic granite (site 34), providing the last chapter in the history of the Gander zone.

Typical Gander zone sediments line the shore of Gander Lake in the Thomas Howe Demonstration Forest, Gander.

# Far Shore
## Sedimentary Rock at Gander Lake

The greenish rock at this site typifies the Gander zone and is the quiet mystery of Newfoundland geology – inconspicuous, enigmatic.

As the Iapetus ocean opened during the Cambrian period, the Humber zone and much of the rest of Canada were part of Laurentia. But the Gander zone lay on the opposite side of the ocean: The rocks here formed from sand and silt washed into the Iapetus ocean on the quiet continental margin of Gondwana.

Where did the sediment come from? What is buried beneath it? Neither of these basic questions has been answered yet. The Gander zone eventually split away from Gondwana, so the source of the sediment is no longer nearby. And in spite of a long search, so far no one has found proven exposures of the continental margin underneath. Even the sediment's age is uncertain. Geologists only know that it formed sometime after the Iapetus ocean opened and before the first stage of its closure (see site 27).

It's an intriguing story – the sedimentary rock here travelled far to become part of Newfoundland, yet little is known about its journey.

# Getting There

## Driving Directions

In Gander, about 3.75 kilometres east of Cooper Boulevard, watch for signs for the Silent Witness Memorial. Turn south onto the gravel access road for the site and follow it about 1.2 kilometres downhill past the memorial to the lakefront.

## Where to Park

Parking Location: N48.91195, W54.57801

This is a gravel parking area along the lakeshore.

## Walking Directions

From the parking area, cross onto the cobble beach and walk northwest along the lakeshore to the outcrop. For an alternative route, look for the trail that leads from the parking area into the adjacent Thomas Howe Demonstration Forest. Follow the trail along the shoreline, then scramble down into the cove to reach the outcrop.

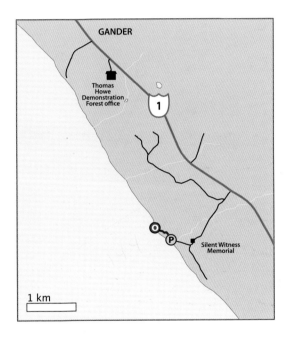

---

**1:50,000 Map**

Gander 002D15

**Provincial Scenic Route**

Kittiwake Coast: Road to the Shore

# On the Outcrop

Quartz veins criss-cross an outcrop of metamorphosed sediment on the north shore of Gander Lake.

Outcrop Location: N48.91324, W54.58085

The outcrops here are metamorphosed sedimentary rock. The original sand and clay were later buried deep enough in the crust to be heated, squeezed, and recrystallized. The greenish colour is due to the presence of chlorite, a common metamorphic mineral. Fluids circulating in the rocks during metamorphism created the numerous quartz veins.

The word most often used by geologists to describe the Gander zone's sedimentary rock is "monotonous," because its appearance varies so little from one outcrop to another. This site is typical of others around Gander Lake and along the Trans-Canada Highway east of Gander.

In addition to the outcrop location noted above, there are many smaller examples all along the shore between the parking location and this site.

1000    900    800    700    600    500

$Z_1$     $Z_2$     $Z_3$     €

# FYI

- Zircon grains survive weathering and travel along with other grains of sand. They preserve information about the age of the rock from which the crystals were eroded. Some zircons found in Gander zone sediments have ages between 1,500 and 1,000 million years. They came from Proterozoic regions of Gondwana.

- The youngest zircons found in Gander zone sediments are about 560 million years old, so the sediment was deposited sometime after that.

- Metamorphism of the sedimentary rock around Gander probably occurred in the Ordovician period, about 465 million years ago, when a slice of Iapetus ocean crust was pushed onto the Gander zone (see site 27).

- Very few fossils are preserved in the Gander zone. The zone was located in cold regions of the southern hemisphere when the sediments were deposited; it is possible that not many life forms inhabited the region.

## Related Outcrops

Additional outcrops of metamorphosed sedimentary rock can be found nearby in the Thomas Howe Demonstration Forest, for example along its Edgar Baird Trail.

The greatest extent of similar rocks is in northeastern Newfoundland; however, additional examples occur in central and southwestern Newfoundland (see site 29).

Metamorphosed sediments of the Gander zone are shaded purple in the map at right.

## Exploring Further

Town of Gander website, www.gandercanada.com. Follow links under the heading Explore Gander for a map of trails in the Thomas Howe Demonstration Forest.

| 500 | | | 400 | | 300 | | 200 | | 100 | | 0 |
|---|---|---|---|---|---|---|---|---|---|---|---|
| Є | O | S | D | C | P | Ṯ | J | K | | Cz | |

Ultramafic rocks (far left) from beneath the Iapetus ocean form the backdrop to an inviting beach at Little Harbour, Gander Lake.

# Ocean Trace
## Serpentinite at Little Harbour

Over a period of several million years during the Ordovician period, the Gander zone was the site of a slow-motion encounter with the Iapetus ocean. An allochthon of ocean crust and underlying mantle from the Gondwanan side of the Iapetus ocean basin was pushed onto the Gander zone.

Whether by coincidence or common cause, this happened about the same time that the Hare Bay and Humber Arm allochthons were pushed from the other side of the Iapetus ocean onto the Humber zone (see sites 6-11).

Evidence of the Iapetus-Gander collision appears as a string of altered mantle rocks along the western edge of the Gander zone in northeastern Newfoundland. The line extends from Ragged Harbour to Gander Lake, with a few additional fragments farther south. In the example at this site, mantle peridotite has been altered to form serpentinite.

# Getting There

## Driving Directions

On the Trans-Canada Highway about 12.5 kilometres east of the bridge in Appleton, turn southeast onto a gravel road at the sign for Little Harbour. Follow the main track of the gravel road for about 2 kilometres, ignoring side roads as you travel downhill toward the lake. At the harbour, bear left and follow the road as it loops around a group of tall trees to a pebbly beach.

## Where to Park

Parking Location: N48.94425, W54.72401

This is an open area adjacent to the small beach.

## Walking Directions

Walk across the pebbly beach to the outcrop location, which is a prominent rock formation where the ground rises behind the beach.

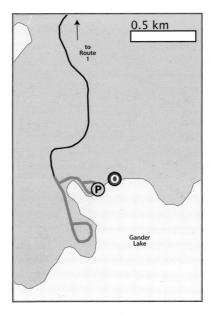

---

**1:50,000 Map**

Gander 002D15

**Provincial Scenic Route**

Kittiwake Coast: Road to the Isles

# On the Outcrop

The serpentinite at Little Harbour on Gander Lake has the waxy lustre and fibrous habit typical of this rock type.

**Outcrop Location: N48.94458, W54.72310**

Geologists spend a lot of time in the field. "The best geologist is the one who's seen the most rocks," some say. This site is an excellent place to experience two properties used in mineral identification: lustre and habit. Serpentine, the dominant mineral in the rock serpentinite, illustrates both well.

Lustre describes the way light interacts with the surface of a mineral. The names for different types of lustre may sound rather old-fashioned, but in practice they can prove useful. Serpentine is described as having a "greasy," "waxy," or "silky" lustre. The fresh surface of the rock in Little Harbour, although hard, looks a bit like a big chunk of crayon, especially on a sunny day. That's waxy lustre.

Habit describes the visible form and growth pattern of a mineral. Serpentine has a "fibrous" habit, in which long, thin, fibre-like crystals are aligned in a dense bundle. Certain surfaces of the outcrop illustrate the fibrous habit well.

In addition to this outcrop, there are other examples of serpentinite around Little Harbour. Get your eye in and take a look.

| 1000 | 900 | 800 | 700 | 600 | 500 |
|---|---|---|---|---|---|
| $Z_1$ | | $Z_2$ | | $Z_3$ | € |

# FYI

- Related events involving the collision of the Gander zone and Iapetus have been recognized farther southwest in the Appalachians of Maine and New Brunswick, where they are known as the Penobscot orogeny.

- The outcrop at this site is part of the Gander River Ultramafic Belt, or GRUB Line (shaded purple and labelled in the map below). Geologists originally thought of the GRUB Line as the western edge of the Gander zone. Later discoveries showed that the Gander zone lies under most, if not all, of the eastern side of the Dunnage zone, too.

## Related Outcrops

Farther west, a string of mantle rocks lies between the Dunnage zone and the Humber zone along the Baie Verte-Brompton Line (shaded purple and labelled BV-BL in the map at right; see sites 18 and 19).

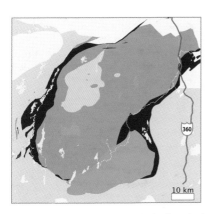

The Tim Horton's structure. Ophiolite, dark purple; metamorphosed Gander sediment, light purple (severely metamorphosed, light pink); granite, dark pink; other Gander and Dunnage zone rocks, dark grey; younger rocks, light grey.

South of Grand Falls is a doughnut-shaped area geologists call the Tim Horton's structure. There, a round area of metamorphosed Gander sediment peeks through the Dunnage zone above. A ring of ophiolite, including mantle rock, marks the boundary, providing evidence for a Dunnage-over-Gander layering of the zones.

A sliver of mantle peridotite in the Tim Horton's structure can be seen in a roadside quarry along Route 360 (Baie d'Espoir highway), just south of the Northwest Gander river (N48.55664, W55.49258).

500    400    300    200    100    0

Є   O   S   D   C   P   Ŧ   J   K   Cz

Layering at Misery Point in Margaree formed during deep burial of a mixture of rock types.

# Tectonic Taffy
## Complex Gneiss at Margaree

As anyone who has made taffy or pastry knows, after a lot of stretching and folding, the end result is very different from the ingredients you started with. The same idea applies in geology; rocks that are layered after being re-worked by heat and tectonic pressure are called gneiss.

Along Newfoundland's southwest coast, gneiss once buried deep in the Earth's crust is now at the surface. Unlike other regions of the island, where original rock features are well preserved, the southwest coast presents a special challenge for geologists trying to understand the history of the region. What were the rocks like before they were buried?

You can't un-pull taffy or un-bake a croissant. But geologists have figured out what happened at Margaree before the gneiss formed: Several kinds of molten rock intruded into the Gander zone in the middle of the Ordovician period as a new section of the Iapetus ocean opened. Later, when Iapetus closed, the intrusions were buried and transformed into the gneiss on view today.

# Getting There

### Driving Directions

From Route 470, follow Route 470-10 (Margaree-Fox Roost Road) about 2.5 kilometres toward the coast. In Margaree, turn left toward Fox Roost. About 600 metres farther, a side road leads to the shore; you may see a small sign for the Misery Point Fishing Livyer's Station. Follow the side road to the shore, about 150 metres.

### Where to Park

Parking Location: N47.57154, W59.05629

This is an open area by the Misery Point Fishing Livyer's Station.

### Walking Directions

From the parking location near the Livyer's Station, follow a gravel path and then a grassy track onto the outcrop, an extensive rock pavement by the shore.

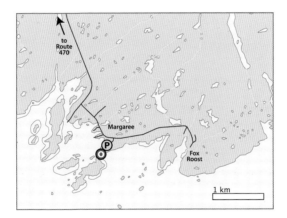

---

**1:50,000 Map**

Port aux Basques 011O11

**Provincial Scenic Route**

Rose Blanche Lighthouse Scenic Drive

# On the Outcrop

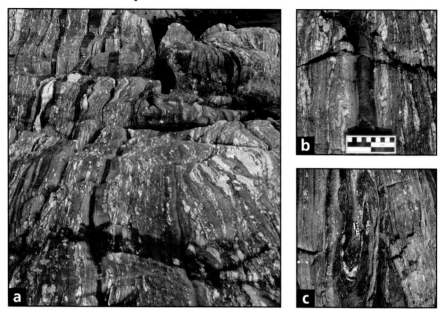

Layering in the gneiss at Margaree takes interesting and attractive forms: (a) Shades of grey hint at a range of original rock types; (b) a narrow layer of dark amphibolite bordered by small folds; (c) layers that have been folded multiple times.

## Outcrop Location: N47.57089, W59.05704

At first glance the gneiss at Misery Point in Margaree may seem like a confused grey mess. But closer study will allow you to identify several points of interest.

The lightest areas on the outcrop are made mainly of quartz and feldspar. When rock is heated, quartz and feldspar are the first minerals to melt. Many of the whitish blobs, strings, and pods in the outcrop have "sweated out" of the gneiss due to high temperature conditions.

At the other end of the colour spectrum there are very dark, almost black bands in the lighter grey. The black rock, called amphibolite, takes many forms and may represent several ages of rock. Some seem to be narrow dykes that cut across other layers and may be slightly younger. Others have been folded into complex shapes, suggesting they were there during the whole process of folding.

The light grey rock is similar in composition to the tonalite at Little Harbour, Twillingate (see site 14), or Tilting, Fogo (see site 23).

1000    900    800    700    600    500

Z₁    Z₂    Z₃    €

# FYI

- Rock chemistry provided many of the clues used to figure out the origin of the Margaree gneiss. Measurements from gneiss samples were compared to measurements from known, unaltered rock types to find the best match, a bit like a chemical fingerprint.

- While the original igneous rock formed during the Ordovician period, it wasn't until continental collision during the Silurian period caused the rock to be deeply buried, folded, and recrystallized to form gneiss.

- Why were the rocks of the southwest coast so deeply buried? It is probably due to the shape of Laurentia and Gondwana as they moved toward one another: Two promontories collided in the southwest, but the rest of Newfoundland lay between reentrants – a sort of gap where the colliding continental margins curved away from one another.

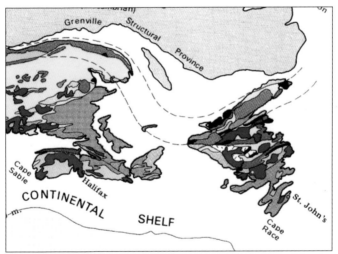

Appalachian Orogen–Tectonic Lithofacies (detail; see page 16 for the complete map).

In the map above, the upper dashed line represents Logan's Line, which defines the western boundary of the Appalachian orogen and runs parallel to the ancient margin of Laurentia. The St. Lawrence promontory is the area where Logan's line curves out toward Cape Breton and southwestern Newfoundland.

## Related Outcrops

The rocks at Margaree intruded sedimentary rock of the Gander zone (see site 26). In southwestern Newfoundland, metamorphism transformed the sedimentary rock into gneiss, with large red garnets and other minerals that form only deep in the Earth (see site 29).

500    400    300    200    100    0

€  O  S  D  C  P  T̶  J  K  Cz

Shoreline outcrops of schist occur along the Grand Bay West Beach Trailway on Granby Point, Port aux Basques.

# Under Pressure

## Garnet-Kyanite Schist at Port aux Basques

For many of the island's visitors, Port aux Basques is the first stop. Visiting rock enthusiasts will be particularly pleased: The Grand Bay West Beach Trailway here provides pleasant access to some interesting examples of schist located along the shores of Granby Point between two beautiful sand beaches.

Schist forms deep in the Earth. There, elevated temperature and pressure, coupled with movement of tectonic plates, cause the growth of metamorphic minerals aligned to form foliation. Schist has finely layered foliation dominated by flat, sheet-like minerals such as mica.

British geologist George Barrow realized in the 1890s that seeing certain minerals together in a rock is like having a thermometer and pressure gauge to measure the metamorphic conditions. Along this trail, garnet and kyanite are found in a schist that formed from sediments of the Gander zone. Garnet and kyanite indicate unusually high temperatures and depths (650°C and 25 kilometres or more below the Earth's surface), a type of metamorphism rarely seen in Newfoundland.

# Getting There

### Driving Directions

From the Trans-Canada Highway near Channel-Port aux Basques, follow Grand Bay Road or Grand Bay Road West to the area immediately west of the main bridge that links the two roads. Turn onto Kyle Lane and follow it for about 250 metres to the parking area.

### Where to Park

Parking Location: N47.58292, W59.18450

This is the parking area for the Grand Bay West Beach Trailway.

### Walking Directions

From the parking area, follow the boardwalk south along the shore. An interpretive panel is located about 0.5 kilometres from the trail head; continue around the headland to the outcrop location.

### Notes

While the trailway is passable in normal tide conditions, the designated outcrop for this site is most easily accessible at low tide. Grand Bay West Beach Trailway passes through the nesting area of the endangered piping plover; the trail is located within the Town of Channel-Port aux Basques Municipal Wetland Stewardship Zone.

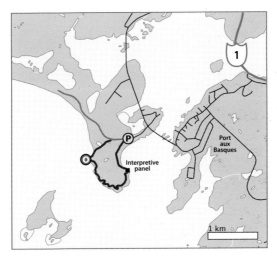

---

**1:50,000 Map**

Port aux Basques 011O11

**Provincial Scenic Route**

Rose Blanche Lighthouse Scenic Drive

# On the Outcrop

Dark schist on the western side of Granby Point contains lighter-coloured pods rich in quartz and feldspar. Near some pods are easily visible clusters of metamorphic minerals.

## Outcrop Location: N47.58064, W59.19188

The outcrop is near where the trailway passes via a boardwalk in front of a small barn, in view of the second (western) beach. The small point of rock, best exposed at low tide, is made of dark-coloured schist.

The most easily visible grains of red garnet and light blue kyanite can be found along with shiny black biotite near irregular pods of light-coloured quartz and feldspar.

The rock here probably began as a sedimentary rock, but the layering you see at this site does not represent layers of sediment. It is due entirely to squeezing and folding that occurred as the schist formed.

This site is similar to the outcrop by the interpretive panel (N47.57937, W59.18455), but at low tide it is more extensive and easily accessed.

Schist (detail).

1000    900    800    700    600    500

Z₁        Z₂        Z₃    €

# FYI

- The rocks at this site are probably equivalent to metasedimentary rock in the northern part of the Gander zone (see site 26). They have a similar composition and occur in the same relative position between the Dunnage and Avalon zones.

- Rocks at this site, once deeply buried, came to the surface through movements along the Cape Ray fault. Rocks east of the fault moved up compared to rocks west of the fault during continental collision after the Iapetus ocean closed.

## Related Outcrops

A hike of about 2 kilometres from J.T. Cheeseman Provincial Park eastward along the provincial T'Railway will take you across the Cape Ray fault. The main fault zone occurs between Windsor Point and the first bridge on the Big Barachois about 800 metres farther east.

In the fault zone, mineral grains have been ground up by movements along the fault. The resulting rock is fine-grained and pink or beige, with strongly defined plate-like layering. Some outcrops include larger grains of partially crushed light-coloured feldspar.

The Cape Ray fault is part of the Red Indian Line (see sites 16 and 21) dividing Laurentia from Gondwana. In other words, the walk along this section of the T'Railway takes you across the former site of the Iapetus ocean.

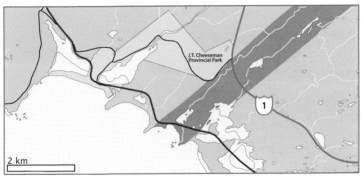

The T'Railway crosses the Cape Ray fault zone (shown in grey) about 2.5 kilometres east of the provincial park.

## Exploring Further

Newfoundland T'Railway Provincial Park website, www.trailway.ca. Follow links at the site for maps and other information about the T'Railway in this region.

500    400    300    200    100    0

Є    O    S    D    C    P    Ŧ    J    K    Cz

The lighthouse at Rose Blanche is a showcase for the Rose Blanche granite, of which it is constructed.

# Dimension Stone

## Syn-Tectonic Granite at Rose Blanche

Author Robert Louis Stevenson knew a thing or two about lighthouses. His grandfather Robert, father, David, uncles Alan and Thomas, and even his cousins were all lighthouse engineers. They provided the original lighting mechanism for the Rose Blanche lighthouse, built in 1871.

Stevenson seemed to understand the significance and spirit of these structures. "There eternal granite hewn from the living isle … stands in the sweep of winds, immovable," he wrote of a favourite beacon in 1887. Those words fit the Rose Blanche lighthouse as well as any: It is built of granite quarried from the cliffs of the headland on which it stands.

The Rose Blanche lighthouse is an excellent place to get a close look at a syn-tectonic granite. It intruded into sediments related to those at Port aux Basques (see site 29) while they were being metamorphosed (heated and folded) deep in the Earth's crust.

Like many others in Newfoundland, the Rose Blanche granite formed as a complex continental collision squeezed and wrenched the Gander and Dunnage zones after the Iapetus ocean closed.

# Getting There

### Driving Directions

From Route 470 near its end point in Rose Blanche, follow road signs for the Rose Blanche lighthouse: Turn left onto Big Bottom Road, then left onto Water Bottom Road, and finally left onto Lighthouse Road itself. Follow Lighthouse Road to a cluster of amenities at the trail head.

### Where to Park

Parking Location: N47.60489, W58.69207

This is an open area in a cluster of tourist amenities around the trail head.

### Walking Directions

Check in at the trail head office to pay your fee, then proceed along one of two gravel trails to the lighthouse. The lower trail includes a coastal section with good exposure of the local rock; the upper trail leads to a lookout with excellent views of the surrounding rocky landscape.

### Notes

The fee includes a tour of the lighthouse, which is a Registered Heritage Structure.

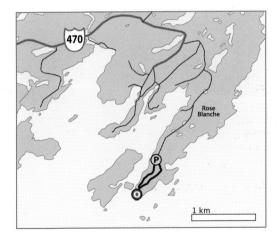

---

**1:50,000 Map**

Rose Blanche 011O10

**Provincial Scenic Route**

Rose Blanche Lighthouse Scenic Drive

# On the Outcrop

Expansive outcrops of granite line the shore along the Rose Blanche lighthouse trail.

**Outcrop Location: N47.60166, W58.69481**

If you take the shoreline path to the lighthouse, then you'll have great views of the Rose Blanche granite for about the second half of your hike. The granite forms big rock pavements along the shore.

At the lighthouse you'll see why its walls are the best place for a close look at the granite. There the rock is well lit and at eye level, ideal for viewing details. The quarried blocks are of medium-grained, very homogeneous rock – overall, there is little variation in the mottled grey colour.

Looking closely, though, in many blocks you'll be able to see that the minerals in the

Rose Blanche granite (detail).

rock are aligned, forming very faint trails of lighter and darker grey. That shows the granite was intruded while the terrain was still being squeezed and folded.

1000  900  800  700  600  500

Z₁  Z₂  Z₃  €

# FYI

- The granite forms large, flat slabs of rock along parts of the shore. They resemble layers of sedimentary rock, but actually they are caused by exfoliation, a complex process that causes cracks to form parallel to the surface of many granites.

- Because the rocks of this area have all been folded together, the granite and its surrounding metasedimentary rock form a complicated pattern of outcrops in the communities of Rose Blanche and nearby Harbour le Cou. If you explore beyond the lighthouse peninsula, you may encounter both types of rock.

The community of Rose Blanche is underlain by granite and metasedimentary rock.

## Related Outcrops

Granitic intrusions of similar age are widespread within the Gander zone (see site 32).

In the map at right, Gander zone sedimentary rock is shown in a lighter hue for comparison; granitic intrusions are darker.

Rock layers at the eastern end of the beach at Windmill Bight Park lie perpendicular to the shore.

# Sheer Depth
## Gneiss and Granite at Windmill Bight

Turkey, Afghanistan, China – these mountainous regions are often in the news because where continents collide to form mountain chains, earthquakes are frequent. Plate tectonic movements cause earthquakes when cold, brittle crust breaks suddenly.

At deeper, hotter levels beneath the mountains, the Earth's crust is weaker. A bit like putty, the crust can change shape rather than breaking. When that happens, instead of a sharp crack along a single fault, there is a longer period of slow, continuous movement along a wide zone of smeared-out rock. That's called a shear zone.

This part of the Gander zone, like the southwestern region (see sites 28-30), has been deep in the Earth's crust. Later, the rocks were pushed up along the Dover fault (see site 33), exposing processes that had taken place far below the surface.

Around the same time that the Iapetus ocean was finally closing along the Dog Bay Line farther west (see site 23), Avalonia drifted to the eastern side of the Gander zone and side-swiped it, so to speak. At Windmill Bight you can examine rocks from a shear zone that formed when that happened.

# Getting There

### Driving Directions

On Route 330 about 2.5 kilometres southeast of Lumsden, watch for the sign at Windmill Bight Park. Turn east onto the park's gravel access road and follow it to the check-in point. Once you have registered, continue along the road to the parking area by the shore. The drive from the highway to the shore is just under 1 kilometre.

### Where to Park

Parking Location: N49.27608, W53.56287

This is the parking area for swimming and beach access.

### Walking Directions

From the parking location, follow any one of a number of paths over the dunes to the beach, then walk to your right (east) along the beach to the outcrop.

### Notes

Windmill Bight Park charges fees for day use and camping.

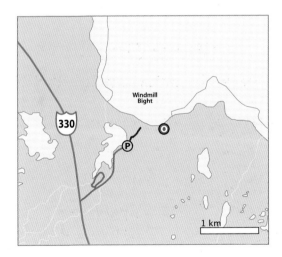

**1:50,000 Map**

Musgrave Harbour 002F05

**Provincial Scenic Route**

Kittiwake Coast: Road to the Shore

# On the Outcrop

The gneiss on the beach at Windmill Bight Park combines layers of metamorphosed sediment with layers of intruded granite.

## Outcrop Location: N49.27766, W53.55706

The gneiss on the eastern end of the beach at Windmill Bight has light and dark bands of several widths. The narrowest bands originated as alternating sand-rich (light) and clay-rich (dark) layers of the original sediment.

Wider light-coloured bands are sheets of granite that intruded the sediments. There are also some dotted layers containing pinkish crystals of feldspar 1 to 2 centimetres across. The crystals have been rounded by the movement in the shear zone. These dotted layers may have originated as granite also.

The shear zone at Windmill Bight is classified as sinistral, meaning left-handed. To picture what this means, stand on the beach and look eastward across the layering toward the high ground. Hold your left arm straight out from your side and point your finger. The far side of the shear zone moved in the direction you are pointing. It dragged and smeared the rocks of the Gander zone into the long, narrow layers you see here.

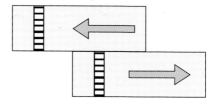

Sinistral movement (map view).

| 1000 | 900 | 800 | 700 | 600 | 500 |
|------|-----|-----|-----|-----|-----|

| Z₁ | Z₂ | Z₃ | € |

# FYI

- The gneiss at this site is equivalent to the sedimentary rock at Gander (see site 26). It looks different because the sediment here was metamorphosed at a much deeper crustal level.

- The rocks at Windmill Bight are part of a structure geologists call the Cape Freels shear zone. It is one of several large shear zones found in a band about 20 kilometres wide along the western side of the Dover fault.

## Related Outcrops

Granites at Greenspond (see site 32) and Cape Freels were intruded into Gander zone gneiss while the Gander and Avalon zones were colliding, and while the shear zone at Windmill Bight was forming.

To view the Cape Freels granite, turn east from Route 330 about 2.5 kilometres southeast of Windmill Bight Park. Follow Cape Freels Road for about 4 kilometres; turn north and drive another kilometre through the community of Cape Freels North. Park at the end of the road and follow the footpath north to the beach; then walk north along the beach.

Granite outcrops are plentiful along North Bill Cove (for example, between N49.26699, W53.49936 and N49.27081, W53.50221). The granite is complex; several kinds of rock surround or cut across one another.

The Cape Freels granite in North Bill Cove, Cape Freels, includes coarse- and fine-grained varieties of granite.

Rounded outcrops of granite shape the landscape around the community of Greenspond.

# Melts in Motion
## Syn-Tectonic Granite at Greenspond

Joseph Beete Jukes, who performed the first geological survey of Newfoundland, arrived in Greenspond on August 5, 1840, for that purpose. Jukes examined the island's complex granitic rocks in some detail. He found the "alternation and passing of one rock into another ... most remarkable."

Geologists of the twenty-first century have other words for the same thing. They describe the rocks of Greenspond as having formed when several varieties of molten rock rose and mingled in a shear zone. For that reason the intrusions are considered syn-tectonic.

Greenspond is part of a broad swath of the Gander zone adjacent to the Dover fault (see site 33) that was affected by movement between the Gander and Avalon zones during the Silurian period. Greenspond is also one of Newfoundland's oldest outports. Boardwalks and trails allow you to explore the island easily, following in the footsteps of J.B. Jukes.

# Getting There

## Driving Directions

Along Route 320, about 14 kilometres northeast of the bridge at Indian Bay, watch for road signs to Greenspond. Turn onto Greenspond Road and follow it about 14 kilometres to the island and town. There, follow Main Street east to Country Lane; turn north onto Country Lane and follow it to the trail head at the top of the rise.

## Where to Park

Parking Location: N49.07108, W53.56494

This is a small open area by the trail head.

## Walking Directions

Follow the boardwalks and trails along the shore for about 1 kilometre to a fork in the path. Take the right fork, which leads to a lookout with a wave-washed outcrop below.

## Notes

Return by retracing your steps, or by bearing left at the fork noted above to continue around the shore and back to town via Meadus Lane or other routes. The community has erected interpretive panels with maps of the island's trails to help you plan your hike.

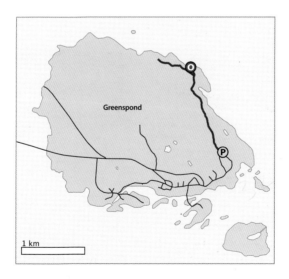

**1:50,000 Map**

Wesleyville 002F04

**Provincial Scenic Route**

Kittiwake Coast: Road to the Shore

# On the Outcrop

Roughly parallel sheets of contrasting rock types are typical of the complex granite in Greenspond.

**Outcrop Location: N49.07877, W53.56915**

There is no shortage of bare rock in Greenspond, but many of the surfaces are darkened by weathering or lichen growth, making it difficult to see interesting details. This outcrop is a lichen-free, wave-washed promontory easily accessed from the trail.

There, you can see the colours and textures that characterize the Greenspond granite – which is not a homogeneous body but a layered collection of rock types. These include a mottled granite with large, clearly visible crystals of feldspar; a pale granite with a fine, sugary texture; and broad, rough sheets of coarse-grained quartz and feldspar.

In contrast to these, you'll see some knobby, dark rafts of coarse-grained amphibolite, as well as large areas of finer-grained grey gneiss cut by a fine web of granitic veins.

Such a collection may seem chaotic, but the overall pattern is of rock types aligned in roughly parallel, upright layers. Geologists call these sheeted dykes. That means there are so many dykes, there is little or nothing between them – just dykes intruding one another. As J.B. Jukes commented, "most remarkable."

1000  900  800  700  600  500

Z₁   Z₂   Z₃   €

# FYI

- The Cape Freels granite (see site 31) occurs in the western half of Greenspond Island and is probably about the same age as the rocks described here.

- Sheeted dykes like those on Greenspond Island form when plate tectonic movements repeatedly open up new spaces into which molten rock flows. At Greenspond, this may have been caused by early movements along the nearby Dover fault.

- The Dover fault (site 33) is not visible on land along this part of the coast, but based on its location and orientation farther south, it probably lies offshore just east of Greenspond.

## Related Outcrops

The Greenspond and Cape Freels intrusions are of similar age to the Rose Blanche granite in southwestern Newfoundland (see site 30). Numerous granites intruded the Gander zone during the Silurian period.

In the map below, Gander zone sediments are shown in a lighter hue for comparison; granitic intrusions are darker.

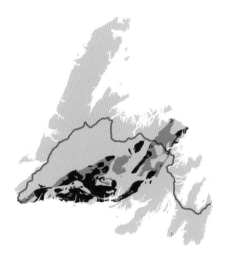

### Exploring Further

Jukes, J.B. *Excursions In and About Newfoundland, During the Years 1839 and 1840.* John Murray, 1842. (Available online at books.google.ca.)

500    400    300    200    100    0

€  O  S  D  C  P  T͞R  J  K  Cz

A series of boardwalks and stairs lead to the lookout in Dover.

# Motion Detector
## Major Fault Trace at Dover

It's a funny thing. To see a structure that cuts down through the Earth's crust, you climb up a hill. But of course you need perspective to see something as large as the Dover fault. It marks the eastern margin of the Gander zone and can be traced for about 90 kilometres between Locker's Bay and the Bay du Nord Wilderness Preserve.

The Dover fault is more than a simple break in the Earth's crust. Long ago it was a plate tectonic boundary, the site of a continental collision. Plate movements first brought the Avalon zone alongside the Gander zone late in the Silurian period (see sites 31 and 32). The Dover fault formed later on, during the Devonian period, as the direction of plate movement changed and the two crustal blocks moved against one another again.

The Town of Dover has made it easy for visitors to get a good view of the fault. Boardwalks and stairs lead to a viewing deck, providing an impressive panorama as well as interpretive panels. Go up – look around – as far as you can see on the clearest of days, the Dover fault reaches farther than that: to the northeast, the southwest, and down into the Earth's crust.

# Getting There

### Driving Directions

On Route 320 a few kilometres north of Hare Bay, watch for signs for Dover. Turn east on Wellington Road and follow it for about 1.75 kilometres into the town of Dover. Watch for signs for the Dover Fault Interpretation Centre on the north side of the road and pull into the centre's parking lot.

### Where to Park

Parking Location: N48.87605, W53.96705

The waypoint marks the entrance to the Dover Fault Interpretation Centre; park in the paved lot in front of the building.

### Walking Directions

Walk around the left side of the interpretation centre and follow the gravel road up a small rise to the start of the boardwalk for the Dover lookout. Then follow the boardwalk to the viewing platform.

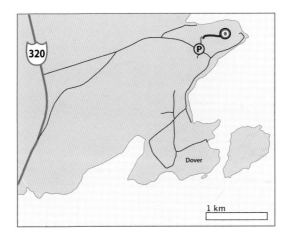

---

**1:50,000 Map**

St. Brendan's 002C13

**Provincial Scenic Route**

Kittiwake Coast: Road to the Shore

# On the Outcrop

The view toward the southwest from Dover lookout: Notches in the landscape follow the trace of the Dover fault.

A fault is a plane of movement in the Earth's crust. Compared to one side of a fault, the other side may go up, down, sideways – whichever way plate tectonic forces are active. Sometimes the line where the fault meets the Earth's surface is visible because the landscape has been affected by past movements. The visible line is called a fault trace.

When movement occurred on the Dover fault, a narrow zone of finely crushed rock formed as the Gander and Avalon crustal blocks ground against one another. Ice age glaciers scoured the region millions of years later, and the weak, crushed rock in the fault zone was easily eroded. A scooped-out notch formed in the landscape.

The fault trace.

At the lookout, the most dramatic view along the fault trace lies back toward the town of Dover. The town itself is in a narrow, low area on the fault trace. Farther away, the beaches by Hare Bay lie in similar low ground.

| 1000 | 900 | 800 | 700 | 600 | 500 |
|------|-----|-----|-----|-----|-----|
| $Z_1$ | | $Z_2$ | | $Z_3$ | $\mathfrak{E}$ |

# FYI

- From the notch at Hare Bay, the fault crosses Freshwater Bay and heads south-southwest toward Maccles, Terra Nova Lake, and Ocean Lake on the boundary of the Bay du Nord Wilderness Preserve.

- Movement along the Dover fault may have been caused by the arrival of yet another Appalachian zone, Meguma. That zone is not seen in Newfoundland but is found in Nova Scotia.

- The Dover fault is a dextral (right-handed) fault. To picture what that means, stand on the lookout facing east-southeast across Content Reach – in other words, looking straight across the fault. Lift your right arm straight out from your side and point your finger. The Avalon zone, on the other side of the fault, was moving in that direction.

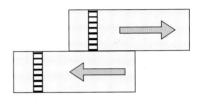

Dextral movement (map view).

## Related Outcrops

The southern extension of the Dover fault is called the Hermitage Bay fault (grey line in the map below). On a clear day you can see its fault trace along Hermitage Bay from the pull-off at the intersection of Routes 360 and 362. Several granites intruded along the line of the Dover fault, after movement along the fault was complete (see site 34).

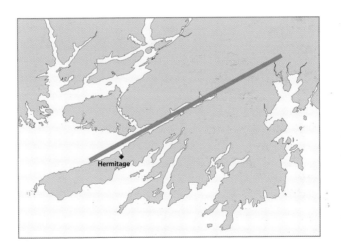

## Exploring Further

Dover Fault Interpretation Centre (N48.87605, W53.96705) is located on Wellington Road in Dover.

"Dover fault." Town of Dover website, townofdover.ca/doverfault.html.

500    400    300    200    100    0

€  O  S  D  C  P  Ṟ  J  K  Cz

The wide expanse of Lumsden beach hosts large outcrops of granite like the one seen here.

# End Game

## Post-Tectonic Granite at Lumsden

A long, complex history led to the Gander zone becoming part of the Appalachian mountain belt. But at Lumsden beach, the story draws to a close. To the west, the Iapetus ocean was gone. To the east, the Avalon zone was attached to Laurentia, and movement had all but ceased on the Dover fault. After 100 million years or more of episodic turmoil, the Gander zone was quiet.

Like signatures on a treaty, a series of granites like this one at Lumsden beach sealed the peace. They cut across previous structures such as gneiss and older granites, and their mineral fabrics contain no evidence of major stress. Because of this, they are called post-tectonic granites.

The outcrops here are part of one of the largest intrusions in the Gander zone. Extending more than 40 kilometres along the shoreline between Musgrave Harbour and the western side of Wild Bight, the granite can be seen at many places along the coast. This site is a good choice: Visitors who walk the broad sweep of Lumsden beach to see these signs of tectonic quiet may find themselves in a state of low stress, too.

# Getting There

### Driving Directions

From Route 330 in Lumsden, follow Lumsden North Road about 1.5 kilometres from the highway. On the right (east) side of the road, just south of the play park, turn right (east) into a sandy track. The parking area is just off Lumsden North Road.

### Where to Park

Parking Location: N49.30920, W53.61497

The parking location is a small open area along a side road to the beach.

### Walking Directions

Follow the sandy track onto the beach, then walk north to the outcrop location.

### Notes

A stream crosses the beach and can be deep at times. If you want to explore further and visit the southern half of the beach, cross the footbridge near the parking location and follow the footpath on the other side of the stream to the beach.

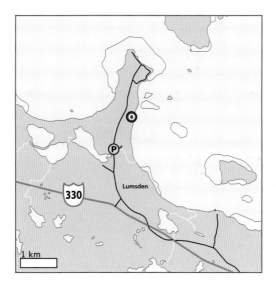

---

**1:50,000 Map**

Musgrave Harbour 002F05

**Provincial Scenic Route**

Kittiwake Coast: Road to the Shore

# On the Outcrop

Staining from a stream that flows across the beach highlights the random distribution of crystals in the granite.

**Outcrop Location: N49.31415, W53.61059**

Wave action has kept the granite outcrops on Lumsden beach well scrubbed, making it easy to see the dramatic, large mineral grains. In one area, a freshwater stream that crosses the beach has stained the outcrop a dark pink, but the rest of the outcrop is unaffected.

Pink crystals of potassium feldspar dominate this rock type. Individual crystals 10 centimetres or more in length are not uncommon. Smaller but still conspicuous are crystals of a white variety of feldspar called plagioclase. The other easily visible minerals are grey quartz and black biotite.

Lumsden granite (detail).

The distribution and orientation of the mineral grains in the outcrop appear random. This is a typical post-tectonic rock texture. In the absence of tectonic stresses, crystals grow freely in any direction.

1000     900     800     700     600     500

$Z_1$      $Z_2$      $Z_3$      €

# FYI

- The granite at Lumsden, sometimes called the Deadman's Bay granite, covers almost 1,000 square kilometres in the northern Gander zone – possibly more, depending on how far it continues offshore.

- The attractive minerals and consistent texture of the Lumsden granite make it suitable for use as building stone. In 1987, a quarry between Lumsden and Windmill Bight provided slabs used in the main foyer of the Alexander Murray Building on Memorial University's St. John's campus.

## Related Outcrops

A similar post-tectonic granite is found in nearby Newtown. If you visit the Barbour Living Heritage Village, look for the building E. & S. Barbour Fisherman's Supplies. Follow a boardwalk around the left-hand side of the building and down a short flight of stairs onto a wooden deck (N49.20158, W53.51533).

A granite outcrop in the Barbour Living Heritage Village.

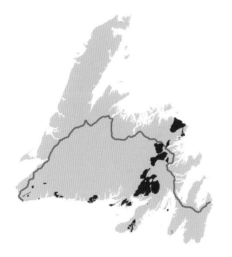

The granites at Lumsden and Newtown are part of a series of granites (shaded purple in the map at left) intruded along and near the Dover fault after significant movement between the Gander and Avalon zones had ceased. (For an example in the Avalon zone, see site 48.)

## Exploring Further

Evans, D.T.W. and W.L. Dickson. *Dimension Stone in Newfoundland and Labrador*. NL Geological Survey Open File NFLD/2865. Government of Newfoundland and Labrador, 2004. (Available online.)

500        400        300        200        100        0

€  O  S  D  C  P  Ṫ  J  K  Cz

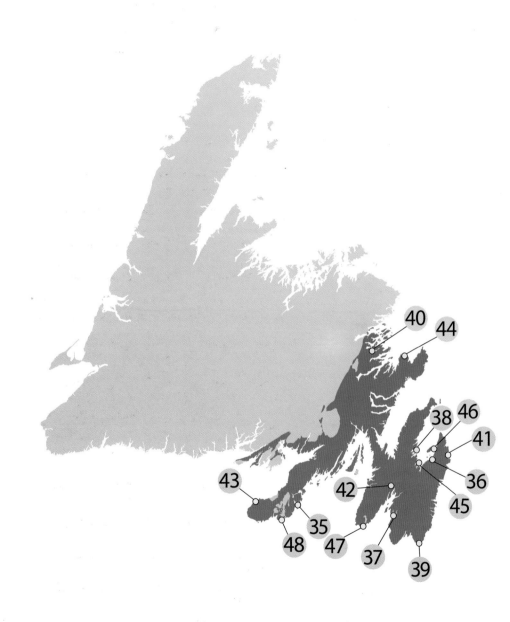

# AVALON

## Zone at a Glance

### Boundaries

West: Dover-Hermitage Bay Fault
East: Not seen in Newfoundland

### Origin

Fragment of Gondwana

### Characteristic Features

Ancient volcanic arcs
Ancient ocean sediment
Rifting, river sediment
Iapetus ocean sediment

## In the Avalon zone, you can …

| | | |
|---|---|---|
| **35** | **Burin** | Visit ocean floor formed near the ancient supercontinent Rodinia. |
| **36** | **Manuels River** | Find the granitic core of the Avalon peninsula. |
| **37** | **St. Mary's** | Encounter Snowball Earth's last glaciation. |
| **38** | **Mad Rock** | Explore a deep basin formed among volcanic islands. |
| **39** | **Mistaken Point** | Experience the dawn of complex life in the ocean. |
| **40** | **Traytown** | Detect a turning point – an unusual granite that signals tectonic change. |
| **41** | **Signal Hill** | Hike through the pages of a geologic drama. |
| **42** | **Cataracts** | Climb down a stairway to view signs of tectonic unrest. |
| **43** | **Fortune Head** | Pinpoint an international boundary of the geologic time scale. |
| **44** | **Keels** | Capture a colourful snapshot of life in the ancient seas. |
| **45** | **Bacon Cove** | Trace the surface of a famous unconformity. |
| **46** | **Bell Island** | Track down some animal tracks preserved in stone. |
| **47** | **Point Lance** | See a rare sill that squeezed between layers of sandstone. |
| **48** | **St. Lawrence** | Inspect our youngest granite, which hosts a fluorspar mine. |

The Avalon zone once formed part of a small continent called Avalonia. Due to plate tectonic movements, the remains of Avalonia are now scattered far and wide and can be found in Newfoundland, the eastern United States, Morocco, western Spain, France, and parts of the United Kingdom.

Even before Avalonia existed, the oldest rocks found in the Avalon zone formed along the margin of the ancient supercontinent Rodinia. Rocks known as the Burin ophiolite formed just as Rodinia began to break up about 760 million years ago (site 35).

Tidal Wave Memorial site (see site 35).

Between 730 and 580 million years ago, a long-lived subduction zone along the margin of Gondwana created a series of granitic intrusions and related volcanic activity. Rocks from this period are now found in an area called the Holyrood horst (site 36).

Volcanic ash and material eroded from volcanic terrains created piles of sediment several kilometres thick in nearby ocean basins. While the sediment was being deposited, the Earth's climate, ocean, and atmosphere were undergoing deep-seated changes. A period of wild climate fluctuations ended, giving rise to an oxygen-rich ocean able to support complex life forms for the first time (sites 37-39).

Mad Rock (see site 38).

Watern Cove (see site 39).

A shift in plate tectonic movements about 570 million years ago led to many changes in the Avalon zone. Intrusions associated with subduction became less important, while intrusions associated with rifting started to appear (site 40).

By 555 million years ago, uplift along a series of major faults had created mountainous areas with low-lying basins in between. Sedimentary layers formed during this time period show that water became more and more shallow as the basins filled in. Eventually rivers flowed across the area, carrying coarser and coarser sediment from the uplifted blocks as tectonic activity continued (sites 41 and 42).

The Avalon zone preserves a detailed record of early complex life forms, including layers formed about 540 million years ago in the Ediacaran-Cambrian Global Stratotype Section and Point at Fortune Head (site 43). A few other small areas preserve fossil shells and tracks left by Cambrian life forms on the margin of Gondwana (sites 44 and 45).

Signal Hill (see site 41).

During the Ordovician period, Avalonia separated from Gondwana and drifted toward the Gander zone (site 46). Very few new rocks formed in the Avalon zone during or after the ensuing continental collision. The rocks of the Avalon zone have never been deeply buried and the youngest intrusions occurred at a shallow level (sites 47 and 48).

The trail to Captain Cook's lookout passes outcrops of the Wandsworth gabbro near Penney's Pond Road in Burin.

# Ocean Fragment
## Newfoundland's Oldest Ocean Crust at Burin

The Burin peninsula is home to both the oldest and youngest rocks in the Avalon zone. Around the community of Burin, the oldest are found; just along the coast in St. Lawrence (see site 48) are some of the youngest. You can easily visit them both in a single day for a whirlwind tour through time.

The rocks at this site formed during the Cryogenian period. All the Earth's landmasses were joined in a single supercontinent geologists call Rodinia (see site 1), and Rodinia was surrounded by a single ocean covering the rest of the globe.

Plate tectonic forces were just starting to pull some regions of Rodinia apart. To make room for the spreading continent, the surrounding ocean began to close – subduction zones formed, creating volcanic arcs in the ocean basin.

Now only fragments of volcanic arcs from that vast, ancient ocean remain. Some are on the Arabian peninsula, some in North Africa. Some have retired to a small quiet pond along Captain Cook's Lookout Historic Trail in Burin.

# Getting There

### Driving Directions

On Route 221 in Burin, about 1 kilometre south of its intersection with Route 221-10 (Harbour View Drive to Port au Bras), watch for signs for Captain Cook's Lookout Trail and for Penney's Pond Road. Follow Penney's Pond Road to the end of the pavement.

### Where to Park

Parking Location: N47.03929, W55.17648

This is a grassy area at the end of the pavement, by the trail head for Captain Cook's Lookout Trail. Please do not block residents' driveways.

### Walking Directions

Follow the gravel track toward the pond and then right, along the east side of the pond. The outcrop is less than 100 metres from the parking location in the hillside beside the pond.

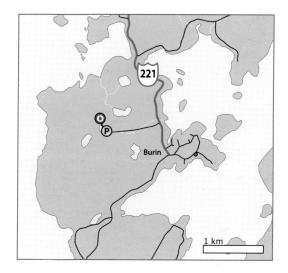

---

**1:50,000 Map**

Marystown 001M03

**Provincial Scenic Route**

Heritage Run

# On the Outcrop

Fine-grained diabase beside the pond on Captain Cook's Lookout Historic Trail is typical of the eastern (lower) edge of the Wandsworth gabbro.

## Outcrop Location: N47.03940, W55.17690

This outcrop is beside the pond along the first part of Captain Cook's Lookout Historic Trail in Burin. The rock here is near the lower edge of a complex mafic intrusion about 1.5 kilometres wide and 30 kilometres long. Known as the Wandsworth gabbro, it intruded between two thick stacks of volcanic rock and sediment, all part of what some geologists call the Burin ophiolite. (Others prefer the name Burin group, since the rocks don't appear to form a complete ophiolite sequence.)

You'll see a variety of rock types between the trail head and the outcrop by the pond. Most share similar mafic compositions, but the size of the mineral grains and the amount of deformation are variable. Some outcrops are part of the intrusion itself, some are xenoliths, and others are mafic schist created by deformation along the margin of the intrusion.

Beside the pond is an even-textured, fine-grained greenish diabase that weathers to a soft brown colour. This is typical of the eastern (lower) margin of the intrusion.

If you continue along the trail to the lookout, you'll encounter related outcrops: The whole trail lies within the Wandsworth gabbro.

$Z_1$    $Z_2$    $Z_3$    $\in$

# FYI

- The Burin Ultramafic Belt (a steep, narrow zone of altered mantle rock) slices through the Wandsworth gabbro. It appears to have been pushed into place during plate tectonic movements. Elsewhere in Newfoundland, younger ultramafic belts can be found near Baie Verte and Gander (see sites 18 and 27).

- The rock chemistry of the gabbro shows that it formed in a primitive volcanic arc, extracted directly from the Earth's mantle. In contrast, most of the younger volcanic rocks in the Avalon zone formed by melting in the Earth's crust. Rock just like the gabbro in Burin may have been the "parent" to much of the Avalon zone.

## Related Outcrops

Outcrops of gabbro, pillow lava, other volcanic rock, and ocean-floor sediment can be seen along most of the walking trails around Burin, as well as:

- Along the harbour boardwalk in Burin. The best outcrops begin near the parking area on Seaview Drive (N47.03671, W55.16297).

- At the Tidal Wave Memorial near Port au Bras. There are outcrops below the lookout point (N47.05638, W55.14417).

The harbour boardwalk in Burin.

The riverbed in Manuels River Linear Park, Conception Bay South, is strewn with loose boulders and outcrops of granite and volcanic rock.

# In the Basement
## Granite and Volcanic Ash at Manuels River

" The gneiss … was first observed immediately above the bridge over Manuel's Brook on the Bay Road … The rock in the bed of the stream is mostly of a pale red or pinkish colour, rather fine grained generally, hard and compact in texture …"

So wrote Alexander Murray, the first director of Newfoundland's geological survey in his "Report for 1868." Today we would call this rock granite rather than gneiss. The bridge, still in the same location, crosses Manuels River on what is now the Conception Bay Highway. But otherwise Murray's account still nicely describes the location and character of the granite in Manuels River Linear Park.

The rocks here are part of a broad spine of mainly granitic intrusions and volcanic rock that extends from the southeastern shores of Conception Bay into the Avalon peninsula's interior. Named the Holyrood horst, the area has been exposed by uplift along faults; it may be an example of the foundation or "basement," as geologists call it, to the many layers of sediment found elsewhere in the region (see sites 37-46).

# Getting There

### Driving Directions

Follow Route 60 in Manuels, Conception Bay South, to a point just west of the Manuels Access Road. Watch for the entrance for the Manuels River Linear Park Chalet.

### Where to Park

Parking Location: N47.52077, W52.94665

This is a gravel parking area beside the Manuels River Chalet. It is accessed from Route 60.

### Walking Directions

From the parking area, walk toward the river and descend the stairway to a large wooden deck. From the deck, follow the path south (upstream) under the highway bridge to the outcrop locations.

### Notes

Manuels River Linear Park is a protected area. No hammering or rock collecting is allowed. The chalet has maps, informative leaflets, and exhibits and is open seasonally.

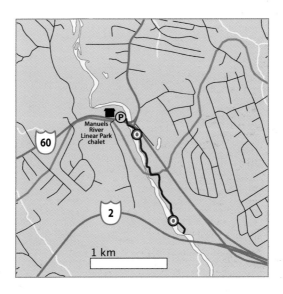

---

**1:50,000 Map**

St. John's 001N10

**Provincial Scenic Route**

Admiral's Coast

# On the Outcrop (1)

An extensive pavement of granite provides a favourite picnic spot between the trail's gazebo and the Route 60 road bridge, Conception Bay South.

## Outcrop Location: N47.51971, W52.94484

This outcrop is a large rock pavement in the riverbed. You can access it from the trail near the gazebo, about 80 metres upstream from the road bridge on Route 60 (Conception Bay Highway).

It's a good place to study the granite because much of the rock surface is clean and free of stains. Pink crystals of potassium feldspar lend the rock its colour. In contrast to the large crystals in younger granites of the Gander zone (see site 34), these are smaller but still plainly visible.

Granite (detail).

The texture and grain size of the granite are quite uniform and there are few signs of deformation. If you explore the outcrop in detail, you may notice the presence of xenoliths (fragments of other types of rock within the granite). Many intrusions contain xenoliths – the fragments fall in or are pulled into the molten rock as it moves through the crust.

1000   900   800   700   600   500

Z₁   Z₂   Z₃   €

# FYI

- Granitic rocks and associated volcanic rocks in the Holyrood horst have ages between about 730 and 585 million years. They all have very similar chemical compositions. It's likely they all formed as part of a volcanic arc near the margins of Rodinia as the supercontinent rifted apart.

- Granitic intrusions of similar chemistry and age also occur along the northeast shore of Conception Bay, between Topsail Head and Cape St. Francis.

## Related Outcrops

About 50 metres farther upstream from the gazebo, the granite comes into contact with an overlying layer of volcanic ash. The contrast between the two rock types is subtle, since they contain many of the same minerals and their colour is similar.

Evidence from other sites in the area shows the ash was there first, and the granite intruded into it. They are very similar in chemical composition and could be about the same age.

The contact between granite (background) and overlying volcanic ash (foreground) is visible as a change from rounded outcrops of granite to more angular, blocky outcrops of ash.

## Exploring Further

Murray, Alexander and James P. Howley. "Report for 1868." *Geological Survey of Newfoundland*. Edward Stanford, 1881. (Available online at books.google.ca.)

# On the Outcrop (2)

Sedimentary rock cut by a mafic dyke along the Manuels River, Conception Bay South.

**Outcrop Location: N47.51282, W52.94029**

From the first outcrop, follow the footpath upstream. It leads up and over the old railway bed toward the Route 2 highway overpass. Fork toward the river when possible to find a variety of outcrops – mainly sedimentary rock that lies above the ash. Some outcrops contain obvious rounded or angular fragments. The sedimentary rock is part of a complex sequence of sedimentary and volcanic layers. They formed during the youngest events in the Holyrood horst.

Within sight of the highway overpass, there is a wooden viewing platform and bench. About 20 metres downstream from the platform, you can step from the path onto an extensive rock pavement in the riverbed. Here, several mafic dykes dissect the sedimentary rock. The dykes vary in width from 10 to 50 centimetres or more. They weather to a soft brown colour, but on surfaces recently broken or worn clean by the river you can see the greenish colour of the fresh rock.

1000      900      800      700      600      500

$Z_1$      $Z_2$      $Z_3$      $\in$

# FYI

- Mafic dykes like the ones seen here may have fed molten rock to volcanoes above – layers of pillow lava and other mafic eruptions are found nearby (though not along the river). The change from granitic to mafic activity about 580 million years ago suggests a change in the movement of tectonic plates as Rodinia broke apart.

- Volcanic ash layers in Conception Bay South are host to two kinds of mineral deposit. Pyrophyllite is an industrial mineral mined in the Oval Pit operations near Manuels. The ash layers also host gold deposits. The economic potential of the gold mineralization is still being investigated.

## Related Outcrops

At the end of the trail, near the waterfall in the area called the Canyon, you again see the older ash layers into which the granite was intruded.

The volcanic ash has a fine-grained, sugary appearance that is quite different from the sediments in the middle part of the trail. Several mafic dykes can be seen cutting across the volcanic ash near the waterfall.

The Canyon area of the park has outcrops of volcanic ash like those near the gazebo downstream.

## Exploring Further

Manuels River Linear Park Chalet is located on Route 60 in Manuels, just west of the road bridge. Visit the chalet for pamphlets, exhibits, maps, and other resources for enjoying the park.

Manuels River Natural Heritage Society website, www.manuelsriver.com. Visit the website for maps as well as updates on the construction of a new interpretive centre for the park.

500    400    300    200    100    0

€  O  S  D  C  P  Ŧ  J  K  Cz

Rock layers containing glacial sediment are exposed along the shores of St. Mary's Bay near the town of St. Mary's.

# Fire and Ice
## Glacial Deposits at St. Mary's

Snowball Earth – Icehouse Earth – Slushball Earth. These are names geologists have given a series of wild climate fluctuations and severe ice ages that gripped the globe during the Cryogenian and Ediacaran periods as the supercontinent Rodinia was breaking up.

On Double Road Point near St. Mary's is a 300-metre-thick band of rock that formed during Snowball Earth's last ice age, called the Gaskiers glaciation. The event was first recognized at this site. Further research showed it was a widespread (though perhaps not worldwide) glaciation that left similar deposits in New England, northern Europe, China, Australia, and South America.

Geologists still argue about what caused ice to cover Rodinia repeatedly even though it lay along the equator. But most agree these events caused profound changes in the ocean. After the Gaskiers glaciation, oxygen became widely available to ocean-dwelling life forms for the first time. The availability of oxygen led to the evolution of Earth's first complex organisms (see sites 38 and 39).

A walk through fields along St. Mary's Bay will lead you to this key moment in Earth history, a turning point for life on Earth.

# Getting There

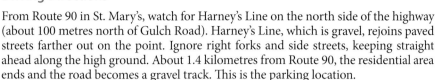

## Driving Directions

From Route 90 in St. Mary's, watch for Harney's Line on the north side of the highway (about 100 metres north of Gulch Road). Harney's Line, which is gravel, rejoins paved streets farther out on the point. Ignore right forks and side streets, keeping straight ahead along the high ground. About 1.4 kilometres from Route 90, the residential area ends and the road becomes a gravel track. This is the parking location.

## Where to Park

Parking Location: N46.92059, W53.58369

Park across the road from the last house, being careful not to block traffic or residents' driveways.

## Walking Directions

From the parking location, walk along the gravel road as it follows the shore, curves back inland, then forks. Select the right fork, which skirts around a fenced-off field and eventually merges with a footpath along the coast. Follow the footpath, then cross a few metres of brushy meadow to the outcrop along the cliff.

## Notes

For alternative access to similar rocks, just north of the bend in Route 90, turn onto Kelly's Lane, then almost immediately turn left, drive about 400 metres, and turn right onto Cove Road. The outcrop is beside the paved ramp at the shore (N46.87921, W53.61256).

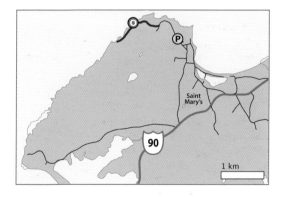

---

**1:50,000 Map**

St. Mary's 001K13

**Provincial Scenic Route**

Irish Loop

# On the Outcrop

Rocks of the Gaskiers formation have characteristics of a debris flow – scattered, diverse fragments in a matrix of finer sediment.

Outcrop Location: N46.92264, W53.59309

The Gaskiers formation, seen at this site, was not deposited directly by glaciers. It was deposited by debris flows when piles of glacial sediment became unstable and slumped down the side of a deep marine basin.

The individual layers are thick (3 metres or more) and most have a massive texture – no bands of coarse or fine grains, no alignment of particles. In the sandy matrix is a mixture of larger rock fragments.

Some fragments have smooth, flat sides or parallel scratches (striations) characteristic of glacial activity. Other fragments were erupted from volcanoes. For that reason these rocks have been referred to as "fire and ice" deposits.

The rock layers here have been tilted by later folding, so as you follow the shoreline path southwest, you are walking across the layers, going back in time.

1000   900   800   700   600   500

Z₁   Z₂   Z₃   €

# FYI

- Prior to the Gaskiers glaciation, the ocean was rich in chemically active iron, sulphur, and organic carbon, but poor in usable oxygen. Such an environment could only support single-celled life forms that in today's world are considered extremophiles.

- Around the world, Snowball Earth episodes appear to have ended abruptly. Sudden melting of the ice contributed to the formation of cap carbonate, a peculiar form of limestone that lies immediately above or "caps" the glacial deposits.

## Related Outcrops

Sediments deposited during the Gaskiers glaciation occur east of Harbour Drive between Harbour Main and Gallows Cove. Debris flow deposits can be seen on Moores Head near the church (N47.44053, W53.15670). Nearby is a small outcrop of cap carbonate.

To see a similar rock in the Bonavista region, follow Route 239 to its termination, then continue on Route 239-15 to Old Bonaventure. Fork left around the east side of the harbour. From a parking and picnic area beside the road, walk about 60 metres farther to a road cut (N48.28507, W53.42554).

Near Harbour Main, the Gaskiers formation includes (a) debris flows containing glacial sediments, and (b) Newfoundland's only known cap carbonate.

## Exploring Further

Attenborough, David. *David Attenborough's First Life.* HarperCollins, 2011. (Chapter 3 reviews the role of Snowball Earth in the evolution of life.)

Snowball Earth website, www.snowballearth.org.

| 500 | | 400 | | 300 | | 200 | | 100 | | 0 |
|---|---|---|---|---|---|---|---|---|---|---|
| Ꞓ | O | S | D | C | P | Ŧ | J | K | Cz | |

The Three Sisters in Spaniard's Bay, a well-known landmark on the Shoreline Heritage Walk near Bay Roberts, are the eroded remnants of a turbidite layer from the Ediacaran period.

# Among the Islands

## Deep-Water Sediments at Mad Rock, Bay Roberts

As you look across Spaniard's Bay from the scenic trail at Mad Rock, the landscape has the comfortable feeling of long-established settlement. The houses of Bishop's Cove are clustered at the base of gentle hills on the opposite shore.

The scene was quite different during the Ediacaran period. In the west, a volcanic arc belched ash into the air. The granitic rock now found in the Holyrood horst (see site 36) formed a ridge to the east. Eruptions and erosion of the volcanic terrain delivered plentiful sediment into an intervening ocean basin.

Mad Rock and the rest of the peninsula at Bay Roberts are part of a group of sediments called the Drook formation that were deposited in such a basin as turbidites. All this happened following the abrupt end of the Gaskiers glaciation (see site 37).

This site has been intensively studied because it was part of the environment in which Ediacaran life forms evolved and lived. Bay Roberts Shoreline Heritage Walk provides easy, scenic access to the area.

# Getting There

## Driving Directions

From Route 70 in Bay Roberts, follow Water Street all the way along the peninsula. Signs for Scenic Mad Rock point the way. The road emerges into an open landscape near the parking location.

## Where to Park

Parking Location: N47.61769, W53.21505

This is a gravel parking area at the end of the paved road. (A gravel road continues from there to additional parking at the point.)

## Walking Directions

From the parking area, follow the gravel track to the point and around to the right as it follows the shore. Just before the track crosses a stream, there is good access to the cobble beach. Walk left onto the beach and double back along it to the outcrop location.

## Notes

The paths at Mad Rock are part of an extensive system of trails in Bay Roberts, the Shoreline Heritage Walk. Maps of the trail system are posted at access points along Water Street.

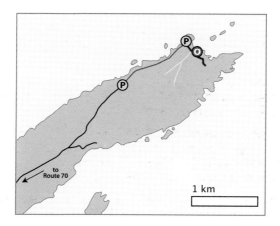

---

**1:50,000 Map**

Harbour Grace 001N11

**Provincial Scenic Route**

Baccalieu Trail

# On the Outcrop

Thick beds of turbidite tilt along the shoreline at Mad Rock near Bay Roberts.

Outcrop Location: N47.62079, W53.20581

The rocks at this site were deposited by weak turbidity currents carrying heavy loads of sediment. The greenish beds are up to 30 centimetres thick and contain fine, parallel bands of silt and mud.

These layers of turbidite are rather monotonous, covering the area all around Mad Rock with similar-looking outcrops. The fact that the beds extend so far without variation suggests they were deposited in a flat region at the bottom of a deep basin.

A few layers have cross-bedding and other sedimentary features that have led geologists to conclude that the turbidity currents carrying the sediment flowed toward the east-southeast.

The rocks are particularly hard, almost flint-like, due to their high content of volcanic ash.

Turbidite (detail).

# FYI

- The earliest large, complex fossils in the world – even older than those in the famous Mistaken Point formation (see site 39) – occur in the upper layers of the Drook formation of the southern Avalon peninsula. (No fossils have yet been found in the Drook formation around Bay Roberts.)

- There are no signs of large, complex life forms on Earth before the end of the Gaskiers glaciation (see site 37), so it is likely that the Drook formation marks a period of very rapid evolution.

- Drook sediments were among the first in the history of the Earth to be deposited in an ocean where deep water was oxygen-rich. This appears to have happened because of changes in ocean chemistry related to Snowball Earth (see site 37).

## Related Outcrops

Turbidites of the Drook formation (shaded green in the map below) are widespread in the Avalon zone, especially along the Irish Loop (Routes 90 and 10). Visitors to Mistaken Point (see site 39) travel across the Drook formation, passing through the settlement of Drook itself on their way to the trail head for a guided tour.

On the northeast Avalon, there are extensive outcrops around Torbay, for example, next to the veterans' memorial across from the post office on Route 20 (N47.66143, W52.73579).

## Exploring Further

Shoreline Heritage Walk, Town of Bay Roberts website, www.bayroberts.com/shoreline.htm. Follow links at the site for a map of the entire trail system.

| 500 | | 400 | | 300 | | 200 | | 100 | | 0 |
|---|---|---|---|---|---|---|---|---|---|---|
| Є | O | S | D | C | P | Ŧ | J | K | Cz | |

Turbidite and ash layers of the Mistaken Point formation, Mistaken Point Ecological Reserve.

# Cellular Revolution

## Ediacaran Fossils at Mistaken Point

Even as his *Origin of Species* was being published in 1859, Charles Darwin remained perplexed. Fossil shells of many kinds of animals had been found in Cambrian rocks, but none at all in older rocks immediately below them. Darwin's own theory suggested the fossilized animals must have evolved gradually from earlier life forms, though none had been found. "If my theory be true, it is indisputable," he wrote, "… that during these vast, yet quite unknown, periods of time, the world swarmed with living creatures."

Darwin was right, as we now know – thanks in part to the discovery of Ediacaran fossils of rare quality and quantity in the Avalon zone of Newfoundland. In the Mistaken Point Ecological Reserve, entire communities of organisms were preserved, confirming the variety and abundance of early life Darwin envisaged.

The site is all the more remarkable because its abundant fossils were all soft-bodied organisms – so primitive no shells had yet evolved. No wonder paleontologists come from all over the world to study the large, gently tilting slabs of sediment exposed on the shoreline between Portugal Cove South and Cape Race.

# Getting There

### Driving Directions

Follow Route 10 in Portugal Cove South, watching for signs to the Edge of Avalon (Mistaken Point) Interpretive Centre on the south side of the road. Check in at the centre; from there, you can choose to drive your own vehicle or ride in the interpreter's van. The drive to the trail head is about 15 kilometres along the unpaved Cape Race Road. (Cape Race is about 6 kilometres beyond the trail head.)

### Where to Park

Parking Location: N46.64135, W53.14067

Park in the gravel area at the trail head. It is marked by an information sign for the Mistaken Point Ecological Reserve.

### Walking Directions

The park interpreter will accompany you from the parking site along a gravel road to the Watern River and from there by footpath to the fossil site.

### Notes

Access to the fossil sites in the Mistaken Point Ecological Reserve is by guided tour only, and space is limited. Contact the interpretive centre ahead of time to make arrangements.

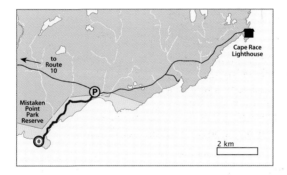

---

**1:50,000 Map**

Trepassey 001K11

**Provincial Scenic Route**

Irish Loop

# On the Outcrop

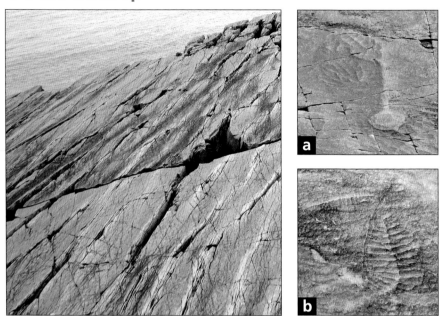

A rock surface on which Ediacaran fossils of Mistaken Point are preserved: (a) *Charniodiscus*; (b) *Fractofusus*.

Outcrop Location: N46.62688, W53.16501

The outcrops are pavements of siltstone at the tops of thick beds of turbidite. The pavement surfaces are rippled and cracked (thought to be the result of rock deformation and later weathering, not ripples formed underwater on the ancient sea bed). In places the dark, gritty remains of an overlying ash layer streak the lighter grey brown of the siltstone.

The fossils that make this locality world-famous are subtle, sometimes elusive imprints of soft-bodied creatures. Best viewed in slanting sunlight but clearly visible in most conditions, they are simply everywhere. You can hardly take a step without treading on them, which is one reason you'll be asked to remove your shoes on the outcrop.

Many of the fossils have round holdfast structures that moored them to the sea bottom. Some of these (photo a) are attached to long slender stalks topped by frond-like shapes. The direction of the stalks on the outcrop is thought to be significant, indicating the direction in which they fell when buried by the ash. Other types of fossils have neither holdfast nor stalk (photo b) and appear to have lived directly on the sea floor.

| 1000 | 900 | 800 | 700 | 600 | 500 |
|---|---|---|---|---|---|
| $Z_1$ | | $Z_2$ | | $Z_3$ | $\in$ |

# FYI

- The collection of creatures preserved on the outcrops at Mistaken Point is a life assemblage – a whole community of organisms in their original life positions. They were quickly buried by volcanic ash like a primordial, watery Pompeii.

- Ediacaran life forms are the earliest large organisms known, appearing rather suddenly at the end of a series of global ice ages (see site 37). The fossil community at Mistaken Point represents a variety of living arrangements – tall organisms on stalks in the same ecosystem as sea-floor dwellers – called ecological tiering. This level of complexity represents another new development in the history of life.

- Although many of the fossils have a plant-like appearance, studies of the sedimentary layers suggest a deep-water environment too dark to support photosynthesis. Based on the large surface area of their frond-like forms, some paleontologists have suggested they absorbed nutrients directly from sea water.

## Related Outcrops

Ediacaran fossils in Newfoundland are restricted to the Avalon zone. Other excellent examples can be found in shoreline outcrops in front of the William Coaker Premises on Main Road east of Route 230 in Port Union (N48.49882, W53.08044).

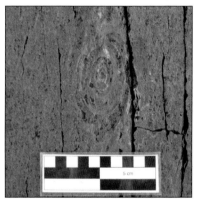

An Ediacaran holdfast preserved along the shore in front of the Coaker Premises, Port Union.

## Exploring Further

Mistaken Point Exhibit. Johnson Geo Centre, Signal Hill Road, St. John's, Newfoundland. A full-sized resin cast of the fossil-bearing rock pavement at Mistaken Point is on display.

Attenborough, David. *David Attenborough's First Life*. HarperCollins, 2011.

Brasier, Martin. *Darwin's Lost World: The Hidden History of Animal Life*. Oxford University Press, 2009.

A small rocky point along Route 310 near Traytown is part of the Louil Hills granite.

# Tectonic Shift
## Louil Hills Granite near Traytown

There's granite on the outskirts of Traytown. Perhaps not so remarkable – 25 per cent of the island of Newfoundland is underlain by some form of granite. That's nearly 28,000 square kilometres of pink stuff. Why come look at this one in particular?

For about 150 million years, the granites and related volcanic rocks of the Avalon zone were very similar in composition. They all had the chemical signature of rock formed above a subduction zone. It's likely the subduction zone lay off the shore of the supercontinent Rodinia, which was in the process of breaking up (see site 1).

Then along came something different, a granite more typically found in a continental rift. For geologists, it signals that tectonic plate interactions were changing. Throughout a long, complicated transition period, there was a gradual shift from subduction to rifting.The changes affected not only the composition of intrusions and volcanic eruptions but also the way sediments accumulated in the Avalon zone (see sites 41 and 42).

The granite forms a small scenic point along Northeast Arm near Traytown. Nearby in Terra Nova National Park, the Louil Hills form the circular core of the intrusion.

# Getting There

### Driving Directions

Follow Route 310 west of Terra Nova National Park. About 1 kilometre east of the bridge to Culls Harbour, watch for a pull-off on the north side of the road.

### Where to Park

Parking Location: N48.65663, W53.94202

This is a pull-off on the north side of Route 310.

### Walking Directions

Walk through a small group of trees to the granite outcrop along the water's edge.

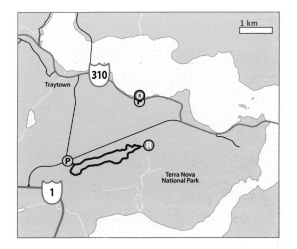

---

**1:50,000 Map**

Eastport 002C12

**Provincial Scenic Route**

Kittiwake Coast: Road to the Beaches

# On the Outcrop

A low outcrop of the Louil Hills granite is exposed along the shore near Traytown.

## Outcrop Location: N48.65673, W53.94194

The outcrop is partly obscured by lichen near the trees, and it is stained brown along the waterline. Between these two zones is an area where you can see its true colours in detail.

Grains of potassium feldspar lend the rock its pink colour. Another type of feldspar (plagioclase) is white; the quartz grains appear grey. The individual crystals are easily visible but not large like those in, for example, the Lumsden granite of the Gander zone (see site 34).

The Louil Hills granite has lower amounts of aluminum than typical granite and because of that it contains no mica (which is aluminum-rich). Instead, it contains riebeckite, a dark blue amphibole.

Granite (detail).

1000    900    800    700    600    500

$Z_1$    $Z_2$    $Z_3$    €

# FYI

- The Louil Hills granite is part of a larger, bimodal intrusion. The mafic portions surround the granite and form a long, oval belt trending north-northeast. (Younger bimodal intrusions can be found in the Dunnage zone, see site 23.)

- The ash layer that covers the fossils at Mistaken Point (see site 39) has a similar age and composition to the Louil Hills granite, although the association is not proven. The search for possible sources of the Mistaken Point ash is still under way.

## Related Outcrops

If you enjoy hiking, you may want to explore the Louil Hills granite further via the Louil Hills Trail in Terra Nova National Park (about a 4-kilometre hike). The hills themselves form the core of the intrusion.

Park at the trail head (N48.64544, W53.96094) and follow the trail to your left, passing through a stand of alders to enter the evergreen forest. Continue along the forest trail to the stairway and ascend to the summit (N48.64789, W53.94027). There you can examine outcrops of the granite and view the neighbouring hills.

From the summit along Louil Hills Trail, the view includes neighbouring hills that also form part of the Louil Hills granitic intrusion.

500          400          300          200          100          0

€   O   S   D   C   P   Ṟ   J   K   Cz

The view from Fort Amherst shows an array of near-vertical sedimentary layers on Signal Hill in St. John's.

# Avalon Upheaval
## Tectonic Basin Fill at Signal Hill

Signal Hill is the undisputed icon of the city of St. John's. A National Historic Site of Canada, it's one of Newfoundland's most visited landmarks, rich in stories of the past.

The rocks of Signal Hill tell stories, too – of the geological past, more than 550 million years ago. At that time the Avalon zone was part of Gondwana. Tectonic forces wrenched the continent, uplifting large blocks of crust and creating basins between the higher regions.

The St. John's area was in such a basin. Rivers flowed southward across the region, depositing layer after layer of sediment eroded from an uplifted block in the north. Continued uplift provided a plentiful supply of sediment; over time an alluvial plain of river sediment formed as the basin filled in.

Later events tilted the sedimentary rock layers on edge, with the oldest in the west and the youngest in the east. The walk from Outer Battery Road to North Head is literally a walk through time. It's a great opportunity to experience what sedimentary rock can tell us about the movement of tectonic plates.

# Getting There

### Driving Directions

From Route 30 in downtown St. John's, turn onto Ordnance Street and then Duckworth Street. Follow Duckworth Street east to the intersection of Signal Hill Road and Battery Road. Enter Battery Road to park.

### Where to Park

Parking Location: N47.57119, W52.69683

This is a suggested location at the west end of Battery Road; alternatively, park nearby as conditions allow.

### Walking Directions

Follow Battery Road, then Outer Battery Road, to the trail head for the Signal Hill National Historic Site. There, an interpretive panel displays a map of the entire trail system.

### Notes

Parks Canada rates this hike as difficult. In adverse weather, check with a Park official for information about trail conditions before starting your hike.

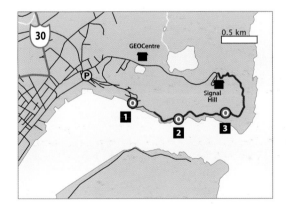

---

**1:50,000 Map**

St. John's 001N10

**Provincial Scenic Route**

St. John's and Environs

# On the Outcrop (1)

The oldest layers on Outer Battery Road are thick beds of homogeneous sandstone: (a) the greenish grey variety; (b) the pink variety, which continues into the National Historic Site.

## Outcrop Location: N47.56907, W52.69114

The outcrop location is near the east end of Outer Battery Road. A small road cut has exposed a fresh surface of greenish grey sandstone with an even texture like medium-grained sandpaper. This layer is near the top of the Gibbett Hill formation. In the past, quarries on Signal Hill supplied blocks of this sandstone for building foundations and bridges in the area.

About 100 metres farther, around a bend in the road, the next outcrop of sandstone is pink. This is the base of the Quidi Vidi formation. As you continue along Outer Battery Road, you'll find several more outcrops, all pink in colour and with the same even, sandy texture.

The two varieties of sandstone contain a similar mixture of sand grains – quartz, feldspar, and tiny rock fragments – so why the difference in colour? Iron in rock turns red only if it is oxidized (like rust). Deep water or rapid burial can choke off oxygen as sediments accumulate, but a reddish tinge means the sediment was exposed to plenty of oxygen as the layers accumulated and hardened. The pink layers on Outer Battery Road are typical of river deposits.

# On the Outcrop (2)

Pea-sized gravel appears near the start of the Signal Hill Trail, Signal Hill National Historic Site.

## Outcrop Location: N47.56775, W52.68604

The next outcrop to note is farther along the trail, about 175 metres from the end of Outer Battery Road. Here the sand is noticeably coarser, and for the first time on this route the sediment includes obvious pea-sized gravel. About 25 metres farther still, you'll see alternating layers of gravel and sand. What was going on here to cause a difference in the size of the sediment?

Changes from one bed to the next can simply be due to changes in how fast the river water flowed. It takes more energy to move gravel than to move sand. For example, gravel may be deposited by fast-moving water in a narrow river channel or during a storm.

As you walk along the trail, you see such fluctuations, but also a trend – a tendency toward more gravel as the layers get younger. When coarser layers are deposited on top of finer layers as a basin fills in, that's called progradation.

Continue along the path to follow the trend.

500    400    300    200    100    0

€  O  S  D  C  P  Ŧ  J  K  Cz

# On the Outcrop (3)

Conglomerate layers containing markedly larger fragments appear between Ross Valley and the stair to the North Head, Signal Hill National Historic Site.

**Outcrop Location: N47.56809, W52.67919**

To find this outcrop, follow the trail past Ross Valley toward the North Head. About 60 metres before the bare ground of the North Head, conspicuous layers of coarse conglomerate appear for the first time in the rock sequence. This rock is typical of the Cuckold formation.

About 80 per cent of the pebbles in the conglomerate are volcanic ash or lava like those in the Holyrood horst (see site 36). Granite pebbles are the next most abundant; a few fragments of sedimentary and dark green mafic rock round out the mix.

As you continue onto the North Head itself, you'll see layers with even larger pebbles. By this point, you've walked straight up through a pile of sediment more than 1.5 kilometres thick – from sand to gravel to cobbles nearly the size of a kiwi fruit. Right to the end, it's a classic example of progradation, that is, sediments becoming coarser over time.

| 1000 | 900 | 800 | 700 | 600 | 500 |
|---|---|---|---|---|---|
| $Z_1$ | | $Z_2$ | | $Z_3$ | $\in$ |

# FYI

- Lots of factors affect the coarseness of sediment. The amount of sediment in the river, the amount of water and how fast it flows, and tectonic uplift nearby can all play a part. Most geologists agree the sedimentary rocks on Signal Hill are syn-tectonic and that the progradation is an effect of the Avalonian orogeny.

- The Gibbett Hill, Quidi Vidi, and Cuckold formations can be traced along the east coast of the Avalon peninsula. The pile of sediment becomes thicker toward the south. Conglomerate clasts in the Cuckold formation are smaller toward the south since they travelled farther and were broken up.

## Related Outcrops

The same layers of rock seen on Signal Hill continue across the Narrows and are exposed along the road to Fort Amherst on the south side of the harbour in St. John's.

Elsewhere in the area, closely related sedimentary rocks occur at the Cape Spear National Historic Site, around Memorial University's Ocean Sciences Centre at Logy Bay, and along the East Coast Trail in the community of Flatrock.

The Cuckold formation conglomerate is found in the easternmost areas of Cape Spear.

## Exploring Further

Geological exhibits and Geo Park, Johnson Geo Centre, 175 Signal Hill Road, St. John's. Indoor exhibits and outdoor interpretive panels provide additional information about the geological history of Signal Hill and other regions of the province.

| 500 | | | 400 | | 300 | | 200 | | 100 | | 0 |
|---|---|---|---|---|---|---|---|---|---|---|---|
| Є | O | S | D | C | P | Ŧ | J | K | Cz |

A waterfall near Ellis Memorial Bridge plunges into the rocky gorge of Ediacaran sandstone at Cataracts Provincial Park near Colinet.

# Silt Story

## Bonavista Basin at the Cataracts

Cataracts Provincial Park is one of Newfoundland's inconspicuous gems. Located on the unpaved Old Placentia highway west of Colinet, the park provides access to a deep, rocky gorge where two rivers converge.

Stairs, bridges, and boardwalks lead into and across the gorge, allowing easy access for exploring its natural beauty and geological features. As an added bonus for fans of bridge architecture, there are excellent views from the gorge of the Ellis Memorial Bridge, completed in 1926 and a classic of that era.

Nature's architect has carved the rocky gorge here through sedimentary layers formed late in the Ediacaran period. At that time, tectonic forces were wrenching the Avalon zone, distorting and rifting the crust along faults to form at least two large basins, one in the east and the other in the west.

The rocks of Signal Hill (see site 41) were forming in the eastern basin while the rocks at this site were forming in the western one. Exact correlations are uncertain; it's possible the rocks here are slightly older.

# Getting There

### Driving Directions

From Route 90 in St. Catherine's near the head of Salmonier Arm, turn west onto Route 91. Follow Route 91 through Colinet, and about 3.5 kilometres west of the intersection with Route 92 watch for the Ellis Memorial Bridge, marked as such by a stone plaque.

### Where to Park

Parking Location: N47.24211, W53.63149

This is a gravel area beside the road, just west of the bridge.

### Walking Directions

Follow the paths, boardwalks, and stairways around, inside, and across the rocky gorge.

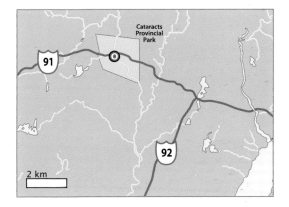

---

**1:50,000 Map**

Placentia 001N04

**Provincial Scenic Route**

Cape Shore

# On the Outcrop

Tilted beds of sandstone emerge from the river at Cataracts Provincial Park.

The outcrop location noted above is at the base of the main stairs into the gorge, where there is a good view of the rock layers in the riverbed. Other outcrops in the riverbed and rocky walls of the gorge can be seen from many points along the bridges, boardwalks, and stairs at the site.

The rock that forms the walls of the gorge is light grey mudstone, a sedimentary rock made of very fine sand. The layers are tilted on edge – they look like big leaning tablets in the riverbed.

Geological reports describe the rock as "wispy-laminated." Along parts of the boardwalk you can get a close look at the mudstone to see what this means. Slight colour variations in the rock highlight the millimetre-scale pulses of sediment, some of them cross-bedded.

Cross-bedding of fine-grained mudstone in the gorge walls shows that the sediment was deposited in shallow, moving water.

| 1000 | 900 | 800 | 700 | 600 | 500 |
|---|---|---|---|---|---|
| Z₁ | | Z₂ | | Z₃ | € |

# FYI

- Zircon grains are commonly found in sand. They survive weathering and erosion, preserving information about where the sediments came from. In a recent study, zircons from sediments related to those here were analyzed. The results showed the sediments were eroded from a terrain like the Holyrood horst (see site 36), with rocks about 600 to 650 million years old.

- Paleomagnetic studies of similar-aged rocks on the Bonavista peninsula and near Argentia show that the Avalon zone was on the margin of Gondwana about 25 degrees south of the equator when the sediments at this site were deposited.

## Related Outcrops

Late in the Ediacaran period, tectonic forces created two basins in the Avalon zone, which then filled with sediment. First deltas and other shallow marine sediments accumulated, followed by river deposits. Sedimentary rock formed in this way (shaded green in the map below) covers much of the Bonavista, Bay de Verde, and Cape St. Mary's peninsulas as well as a long narrow strip along the east coast.

A stairway leads into the gorge at Cataracts Provincial Park.

500　　　　　400　　　　　300　　　　　200　　　　　100　　　　0

€　O　S　D　C　P　Ṟ　J　K　Cz

Ediacaran and Cambrian sediments are exposed along the shore in Fortune Head Ecological Reserve.

# Global Standard
## Ediacaran-Cambrian Boundary at Fortune Head

Until the mid-1960s, generations of geology students learned that the appearance of trilobite fossils marked the beginning of the Cambrian period and the start of life on Earth. The discovery of large, multi-celled life forms 25 million years older than trilobites (see site 39) threw the definition of the Cambrian period into question.

An international working group of scientists was formed in 1972 to select a Global Stratotype Section and Point (GSSP) for the base of the Cambrian period. A GSSP is an official international reference point for the geologic time scale.

Traditionally, the start of each geologic period corresponds to an abrupt, worldwide change in the fossil record. Years of research went into the hunt for an ideal example. Finally, after rejecting sites in Siberia and China, in 1992 the group selected Fortune Head.

At this site, sediment accumulated for millions of years without interruption along the margin of Gondwana. A rich variety of trace fossils records a sudden increase in the number of life forms. The rock layer in which the increase first occurs marks the official start of the Cambrian period. The layers below are the very last record of the Ediacaran period at the end of the Proterozoic eon (see Geologic Time, pages 8 and 9).

# Getting There

### Driving Directions

Follow Route 220 to a point about 2.5 kilometres south of Fortune's town centre. Watch for a large monument-style sign for the Fortune Head Ecological Reserve and turn as indicated. Follow the reserve's gravel access road to the open area near the lighthouse.

### Where to Park

Parking Location: N47.07390, W55.85899

This is a gravel parking area just uphill from the lighthouse.

### Walking Directions

Walk back along the reserve access road, watching for a gravel track on the left. Follow the track through a grassy area and up the hill. Continue along the high ground to the overview location.

### Notes

Fortune Head Ecological Reserve and its fossils are protected by law. It is illegal to disturb or remove any fossils, rock materials, or other natural features.

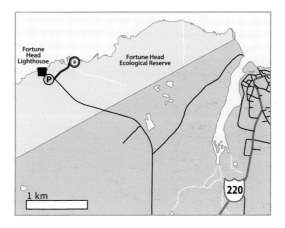

---

**1:50,000 Map**

Grand Bank 001M04

**Provincial Scenic Route**

Heritage Run

# On the Outcrop

The Ediacaran-Cambrian boundary lies within a small rocky point, seen here from the nearby hillside in Fortune Head Ecological Reserve.

The outcrop location is on a hill east of the lighthouse at Fortune Head. It provides a vantage point from which the GSSP boundary layer can be viewed.

The rocks at the site are greenish gray sedimentary rocks, mostly siltstone and mudstone. Late in the Ediacaran period and early in the Cambrian, one or more rivers washed volumes of fine sediment into the sea here. The Ediacaran layers were deposited in shallow water by tidal currents, for example, in a shallow bay. The Cambrian layers were deposited in a similar shallow environment, but nearer the mouth of a river, on a delta.

Many of the layers show signs this was a stormy region. Sedimentary rock deposited

Thin beds of grey tempestite are common along the shore on either side of the lighthouse at Fortune Head.

in stormy conditions is called tempestite. Outcrops of tempestite can be found along much of the shore within the ecological reserve.

| 1000 | 900 | 800 | 700 | 600 | 500 |
|------|-----|-----|-----|-----|-----|
| Z₁ | | Z₂ | | Z₃ | € |

# FYI

- The only trace fossils in the Ediacaran layers at Fortune Head are very simple horizontal feeding burrows. Trace fossils in the Cambrian layers include several kinds of complex feeding burrows, dwelling burrows, and arthropod scratch marks. The specific trace fossil used to define the boundary is a complex burrow, *Phycodes pedum*.

- Tilting of the rock layers means that as you walk west at Fortune Head, you are passing younger and younger layers, as if climbing higher and higher in what was once a vertical stack of rock layers. If they were vertical today, you would have to climb 400 metres above the lowest Cambrian layer before coming to the first shell fossil. You would have to climb 1,000 metres farther before coming to the first trilobite fossil.

- Underneath the Ediacaran rocks of Fortune Head are layers of red sandstone and conglomerate equivalent to the rocks at Signal Hill (see site 42). They are not exposed at this site.

## Related Outcrops

Cambrian and related Ordovician sedimentary rocks (shaded green in the map at right) are uncommon in the Avalon zone. They are only exposed in a few small areas, for example, Conception Bay, Cape St. Mary's, and the Burin peninsula. Unlike the ones at this site, most lie above an eroded surface, or unconformity.

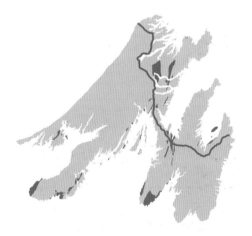

Newfoundland is also the site for the Cambrian-Ordovician GSSP, located at Green Point in Gros Morne National Park (see site 3).

## Exploring Further

Fortune Head Interpretation Centre, 49-51 Bunker's Hill Road, Fortune (N47.07406, W55.82891). The centre includes exhibits related to geologic time. There is a fee to view the exhibits.

Attenborough, David. *David Attenborough's First Life*. HarperCollins, 2011.

Brasier, Martin. *Darwin's Lost World: The Hidden History of Animal Life*. Oxford University Press, 2009.

International Commission on Stratigraphy website, www.stratigraphy.org.

| 500 | | 400 | | 300 | | 200 | | 100 | | 0 |
|---|---|---|---|---|---|---|---|---|---|---|
| Є | O | S | D | C | P | Ṟ | J | K | Cz | |

Rocks formed early in the Cambrian period outcrop around the harbour at Keels. Sandstone in the distant grey cliffs is the oldest; younger slates, mudstones, and limestones can be seen in the foreground.

# Rising Tide

## Early Cambrian Sediments at Keels

One of several small areas of Cambrian rock preserved in the Avalon zone is located in the community of Keels. Most geologists agree that while these rocks formed the Avalon zone was still part of the supercontinent Gondwana.

On the west side of the harbour in Keels is a colourful remnant of early life in the sea. The red and light grey rocks here are closely related to the layers above the unconformity at Bacon Cove (see site 44). Both contain stromatolites and fossil worms.

The layers of sedimentary rock around Keels formed during a time when sea level was rising around the world. First sand, then finer sediment and limestone were deposited here as the ocean invaded the margins of the continent.

Keels has a variety of other geological attractions, the best known being "devil's footprints" (see FYI). The town is also known for the colourful red and green slate quarried nearby. Some of the older buildings in Keels are set on foundations built of this local stone.

# Getting There

### Driving Directions

In King's Cove along Route 235, watch for signs to Duntara and Keels. Turn north onto Route 235-20 (Keels Road) and follow it northwestward to the town of Keels. Continue west around the cove to the parking location; you may see signs for the Devil's Footprints along the same road.

### Where to Park

Parking Location: N48.60533, W53.40760

This is a grassy space beside the road. Please do not block traffic or residents' driveway access.

### Walking Directions

From the parking location, walk west along the road for about 50 metres. As you approach a rocky outcrop on the right, you'll see a path through the grass, leading to the beach below. Descend to the beach and follow the shore around to the west side of the small cove. Make your way onto the outcrop.

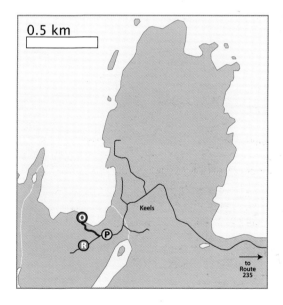

**1:50,000 Map**

Bonavista 002C11

**Provincial Scenic Route**

Discovery Trail

# On the Outcrop

The cracked surface of a grey algal mat lies under a red mudstone on the west side of the harbour at Keels.

Outcrop Location: N48.60608, W53.40927

The sedimentary layers in this part of Keels slant gently down toward the east. Some layers are grey and others are brick red. Don't be distracted by the upright foliation along which the rock breaks so easily. That formed later, when the layers were folded.

The layer with the most outlandish appearance here is a fossilized algal mat. Its top surface looks like a dried-up mud puddle. Exposure to air cracked the mat, and you can see the curled edges of the individual segments.

Above the algal mat is a layer of brick red mudstone. The sediment filtered down into the damaged algal mat and filled in the cracks. Other layers in the outcrop include a light grey limestone that formed as a mound of limestone mud, possibly trapped or secreted by algae or bacteria.

Worm fossils can be found in the rocks at this site. They are long, slender cone-shaped shells up to 15 centimetres long.

1000     900     800     700     600     500

$Z_1$      $Z_2$      $Z_3$      €

# FYI

- In the Avalon zone, there were two phases of sedimentary rock formation during the Cambrian period. The first phase, seen at Fortune Head (see site 43), was focused in the western part of the zone. It was followed by uplift and erosion in some locations. The second phase was focused around Placentia Bay and here on the Bonavista peninsula, including this site.

- On the other side of the Iapetus ocean, deep stacks of sediment also accumulated in the Humber zone early in the Cambrian period. For example, the rocks of Barr'd Harbour's North Summit (see site 1) formed about 520 million years ago.

## Related Outcrops

While in Keels you can visit the landmark "devil's footprints." Beginning from the parking location, continue west along the unpaved road for about 150 metres, until you see a high rock wall on your right (the north side of the road).

The hollows in the rock formed by weathering of easily dissolved limestone (calcium carbonate) concretions in the mudstone. Formation of the concretions was probably triggered by bacteria living in the sediment while it was still soft and wet, buried at the bottom of the sea. The bacteria gave off carbon dioxide, some of which created the "carbonate" part of calcium carbonate. The calcium came from sea water still trapped in the muddy sediment layer.

The holes are arranged in parallel rows because the concretions formed only in specific layers of the rock.

Regularly spaced hollows mark the location of limestone concretions within layers of sediment at Keels.

| 500 | | 400 | | 300 | | 200 | | 100 | | 0 |
|---|---|---|---|---|---|---|---|---|---|---|
| Є | O | S | D | C | P | Ŧ | J | K | Cz | |

Flat-lying Cambrian sediments (lower left) overlie tilted Ediacaran layers at Bacon Cove, Conception Bay.

# Mind the Gap

## Unconformity at Bacon Cove

Alexander Murray, the first director of Newfoundland's Geological Survey, visited Bacon Cove and nearby outcrops in 1867, and reported: "The limestones and red shales may be seen butting against the corrugated edges of the older formation." The significance was not lost on him: "It thus became obvious," he wrote, "there was … a vast difference in age between them."

This quiet cove near the head of Conception Bay is the site of an angular unconformity – a gap in the geologic record where older, tilted rocks are overlain by much younger, less deformed ones. The tilting means the older rocks must have been folded, uplifted, eroded, and submerged before the younger ones were deposited on top. So where mountains once stood, now there's just … the gap.

The surface of the older, tilted rocks is uneven – with obvious hollows, hummocks, and cracks. By strange coincidence, it was a rocky shoreline very similar to the one on which it is exposed at the present day.

# Getting There

### Driving Directions

Along Route 60 between Avondale and Colliers, turn northeast onto the road to Kitchuses and Bacon Cove. Follow the road for about 2.4 kilometres to a fork. At the fork, bear left over the hill and continue about 4 kilometres farther.

### Where to Park

Parking Location: N47.48622, W53.16726

Park on the west side of the road just past a guardrail on the east side. There is room for a car to pull straight in.

### Walking Directions

Cross the road and at the north end of the guardrail enter the field. Make your way through the grass to a small beach at the bottom of the hill. Follow the rock outcrops southeast around the shoreline for about 100 metres.

### Notes

Because of its significance to the geological history of the Avalon zone, Bacon Cove is a protected site. It is illegal to disturb, damage, or remove any rocks found here.

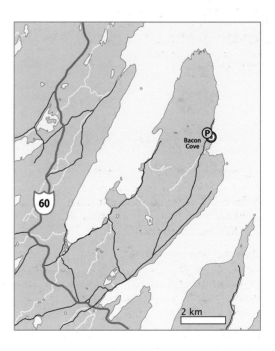

---

**1:50,000 Map**

Holyrood 001N06

**Provincial Scenic Route**

Admiral's Coast

# On the Outcrop

In some parts of the outcrop, Cambrian sediment includes a lower layer of conglomerate. In other parts, stromatolitic limestone rests directly on the unconformity: (a) conglomerate filling a depression, with limestone above; (b) stromatolitic limestone directly above the unconformity.

## Outcrop Location: N47.48520, W53.16613

The rock pavement nearest the parking area is a smoothly worn, red mudstone. It is one of the rock layers above the unconformity.

Farther along the shoreline, fragmented greenish grey shale appears at the surface. It is the older rock below the unconformity. You'll find yourself alternately walking above and below the unconformity on your way to the outcrop location.

At the outcrop there are several places where small vertical rock walls provide excellent views across the unconformity. In some, depressions in the underlying shale are filled with a rocky mixture, including fragments of the shale itself.

Elsewhere a dark, granular limestone lies directly on the unconformity and fills crevices in the rocks below it. Algal mat (stromatolite) structures in the limestone include wavy layering and small mounds.

Below the unconformity it is obvious the rock layers are not horizontal, but they are so broken up it can be hard to see the tilted layering clearly. The best angle for viewing this will depend on light conditions. Outcrops near where you first entered the site provide the best examples of tilted beds below the unconformity. Take a look on your way back to the parking location.

1000    900    800    700    600    500

$Z_1$    $Z_2$    $Z_3$    $\in$

# FYI

- The lower, tilted layers are deep-basin Ediacaran sediments related to those at Mad Rock and Mistaken Point (see sites 38 and 39).

- Deformation of the Ediacaran rocks occurred during the Avalonian orogeny. Related tectonic events affected parts of South America, Africa, Australia, and Antarctica during the same time period.

## Related Outcrops

Another Ediacaran-Cambrian unconformity is exposed in the Manuels river. There, a Cambrian conglomerate lies above the 620-million-year-old granite of the Holyrood horst (see site 36). In other areas of the Avalon, for example, at Fortune Head (see site 43), the rock record is continuous across the Ediacaran-Cambrian boundary.

The Manuels river flows across the large pebbles of a Cambrian conglomerate lying unconformably on Ediacaran granite in Manuels River Linear Park.

## Exploring Further

Geological exhibits and Geo Park, Johnson Geo Centre, 175 Signal Hill Road, St. John's. The exhibits include a large polished slab of the unconformity at Bacon Cove.

Manuels River Natural Heritage Society website, www.manuelsriver.com. Visit the website for maps and information about the park as well as updates on the construction of a new interpretive centre for the park.

Murrray, Alexander and James P. Howley. "Report for 1868." *Geological Survey of Newfoundland*. Edward Stanford, 1881. (Available online at books.google.ca.)

500    400    300    200    100    0

€  O  S  D  C  P  Ŧ  J  K  Cz

Layers of Ordovician siltstone and shale on Bell Island tilt slightly westward by the Beach on the north end of Bell Island Tickle.

# Farewell, Gondwana

## Youngest Avalon Sediments on Bell Island

The ferry leaves the dock in Portugal Cove – you're off to Bell Island. In the distance, the island rides high like an oversized craft in the wide expanse of Conception Bay. Departure and travel are fitting themes for Bell Island's geological past. As you view the cliffs of Bell Island, consider their story:

The rock layers of the island span a significant period in the history of the Avalon zone. When the oldest layers (along Bell Island Tickle) formed, this part of Newfoundland was still part of the supercontinent Gondwana. But by the time the youngest layers (on the bay side) formed, continental rifting had wrenched Gondwana's margin apart to open the Rheic ocean. On one side of the new ocean, a small, new continent, Avalonia, drifted alone.

There are several features of geological interest on Bell Island. The rocks here formed in shallow water during the Ordovician period, so there are many signs of abundant animal life in the form of trace fossils – tracks and burrows. Also, don't miss the mine tour. It's a unique opportunity to walk around inside a rock layer and learn about mining history.

# Getting There

### Driving Directions

From the ferry terminal, instead of going up the hill, turn right and drive eastward toward the small cobble beach below the cliff. Continue across the open gravel area at the end of the road.

### Where to Park

Parking Location: N47.63265, W52.91938

This is an open gravel area near the Beach.

### Walking Directions

From the parking location, follow a trail along the shore, crossing down onto the cobble-strewn shoreline where convenient. Continue along the shore to the outcrop location, about 100 metres from the parking site.

### Notes

Locations of other points of interest on the island can be found in the FYI section.

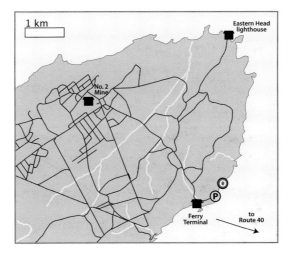

---

**1:50,000 Map**

St. John's 001N10

**Provincial Scenic Route**

Killick Coast Trail

# On the Outcrop

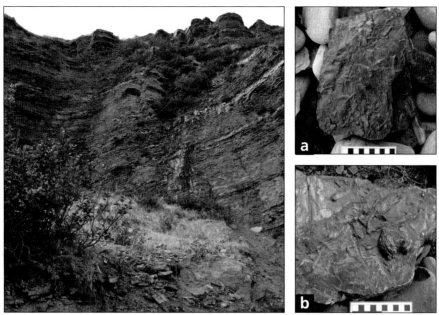

Thin layers of fine sandstone and shale at the Beach near the ferry terminal on the northeast shore of Bell Island yield a wide variety of trace fossils (a and b).

## Outcrop Location: N47.63346, W52.91887

The area called the Beach is surrounded by high cliffs of fine sandstone and shale. The fine sand contains flakes of mica, which makes the rock surfaces sparkle in certain angles of light.

Along the top of the cobble beach, broken layers of dark, rusty brown siltstone are being eroded at the base of the grassy hillside. The siltstone fragments tumble down onto the grey, rounded beach cobbles – the contrast is obvious, and it's easy to pick through the fragments in search of trace fossils.

Trace fossils don't record the shapes of shells or of creatures themselves (as seen, for example, at site 4 or site 39). Instead, they record signs of activity such as tracks and burrows. The shape, size, and pattern of many trace fossils are very specific and can be used just like fossil shells to study the history of animal life and correlate among rock layers. The study of animal traces is called ichnology.

1000     900     800     700     600     500

$Z_1$          $Z_2$          $Z_3$    €

# FYI

- The layers on Bell Island are tilted about 8 to 10 degrees toward the west. For that reason, as you cross the island from east to west you encounter younger and younger layers.

- The iron ore once mined here is found in a single rock layer several metres thick. The mine follows the layer as it dips underground and even under the bay. The ore consists mainly of sand-sized, iron-rich pellets called oolites that were carried by tides or storm currents and deposited in a series of sandbars.

## Related Outcrops

Near the lighthouse at the northeast end of Bell Island (N47.65604, W52.91658), walking trails lead to many scenic views of Eastern Head. These rock layers were the last layers in the Avalon zone to form before it separated from Gondwana. From the lighthouse westward to Redmonds Head and the eastern side of Freshwater Cove are thick layers of sandstone deposited in a rift valley or narrow seaway as the continent broke apart.

The sandstones and shales of Eastern Head, Bell Island, can be viewed near the lighthouse.

## Exploring Further

Bell Island Community Museum and No. 2 Mine Tour, Compressor Hill, Wabana (N47.64645, W52.94945). The museum offers exhibits and guided underground tours of the former mine.

Martin, Wendy. *Once Upon a Mine: Story of Pre-Confederation Mines on the Island of Newfoundland.* Special Volume no. 26. Quebec: Canadian Institute of Mining and Metallurgy, 1983. (Chapter V outlines the history of iron mining on Bell Island.)

Hillside outcrops of diabase line a gravel road along the shore road south of Point Lance.

# Singular Event
## Diabase Sill at Point Lance

Add this to your list of things to do on the Cape St. Mary's peninsula: Visit Point Lance. Travellers to the nearby seabird sanctuary might see a rare bird among the thousands of gannets and kittiwakes. In Point Lance you're assured of a rare sighting – you'll find outcrops of the only known Silurian-age rock formation in the Avalon zone.

The outcrop is part of a Silurian sill, one of several in the area. A sill is an igneous intrusion that spreads out horizontally between existing layers of rock. The sills around Point Lance intruded between layers of Cambrian sandstone and shale.

The Silurian period was an eventful time in other areas of Newfoundland. Parts of the Iapetus ocean were closing, creating mountainous terrains, volcanic activity, and large volumes of sediment (see sites 21 and 22). Not so in the Avalon zone. The microcontinent Avalonia approached the Gander zone without major upheaval. No one yet knows why the sill was intruded here.

# Getting There

## Driving Directions

On Route 100 about 4.5 kilometres west of Branch, watch for signs to Point Lance. Turn onto Route 100-17 (Point Lance Road) and follow it for about 11 kilometres. Continue through the community of Point Lance to the west end of the beach.

## Where to Park

Parking Location: N46.80977, W54.08934

This is a gravel area near the end of the road at the west end of the beach.

## Walking Directions

From the parking location, the road continues along the shore as a gravel track. Follow the track up the hill for about 200 metres to the outcrop.

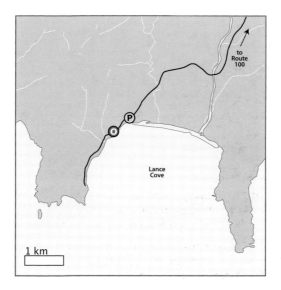

---

**1:50,000 Map**

St. Bride's 001L16

**Provincial Scenic Route**

Cape Shore

# On the Outcrop

The uniform texture and speckled appearance of the diabase makes it easy to recognize.

The outcrop is a mafic rock called diabase. About 30 metres farther along the gravel road there is a second, similar outcrop. (Some parts of this outcrop and others in the area are coarse grained enough to be called gabbro rather than diabase.)

The outcrops contain white feldspar plus a dark mineral called pyroxene. The individual mineral grains are small but visible. If you inspect the rock closely, you'll see the minerals are randomly arranged with no sign of foliation. The absence of any parallel alignment shows the diabase was intruded at a time free of tectonic stress.

Diabase (detail).

Because it was intruded between sedimentary layers, the sill is like a big, flat tablet. Along this road, the sill tilts down toward the road at an angle of about 45 degrees.

From the outcrop there are sweeping views back toward the beach and town. Note that the hill beside the town is another sill.

| 1000 | 900 | 800 | 700 | 600 | 500 |
|---|---|---|---|---|---|
| $Z_1$ | | $Z_2$ | | $Z_3$ | € |

# FYI

- The intrusion divided into five main sills as it flowed among the layers of sediment. The lowest one is the thickest and most extensive. It forms the big headland at Point Lance and extends northward past Route 100.

- After the sills in this region spread out between flat, horizontal layers of sediment, the sills and the sedimentary rock were all folded during the collision between the Avalon and Gander zones, probably sometime during the Devonian period.

- The sedimentary layers around Point Lance are similar in age to those in the downstream portion of Manuels River Linear Park (see site 36).

## Related Outcrops

The track continues from this site along the west side of Lance Cove all the way to Bull Point, crossing several sills along the way. By the abandoned wharf at the Point Lance Tilts (a former fishing stage near Bull Point) is an unusually thick sill with internal layering that formed as the sill cooled.

Most of Newfoundland's Silurian and Devonian mafic intrusions (shaded green in the map below) are located in the Dunnage zone. Some of them occur with granitic rock as part of a bimodal intrusion (see site 23).

500    400    300    200    100    0

€  O  S  D  C  P  Ŧ  J  K  Cz

Granite outcrops and debris create a rose-coloured shoreline east of Chambers Point, St. Lawrence. The granite headland known as Hares Ears is visible in the distance.

# Latecomer

## Post-Tectonic Granite at St. Lawrence

"A is for the axes we use cutting props," begins the "Miner's Alphabet," a traditional song used in the exhibits at the Miner's Memorial Museum in St. Lawrence. Throughout much of the twentieth century, extraction of fluorspar from the nearby granite shaped the town's economy and the lives of its families. Sadly, exposure to rock dust and radon gas cost many miners their lives. The museum commemorates their suffering and celebrates the history of their hard work.

Prior to the twentieth century, the fluorspar itself (which is used in steelmaking and other industrial processes) was of little interest. Early mining operations focused on small amounts of associated lead ore. The value of fluorite increased in the 1920s, leading to a flurry of prospecting activity. Commercial mining began in 1930.

The granite that hosts the fluorspar veins is significant geologically as well as economically. It's the youngest granite in the Avalon zone and one of the youngest in Newfoundland. By the time it intruded the Ediacaran and Cambrian rocks of the Burin peninsula, all four of Newfoundland's geologic zones were united and the newly assembled continent was stable.

# Getting There

## Driving Directions

From Route 220 in St. Lawrence, turn south onto Laurentian Drive and follow it to its end point at Iron Springs Road (named Director Drive on some online map services). Turn right on Iron Springs Road and follow it southwest for about 3.7 kilometres to the well-marked trail head for the Chambers Cove Hiking Trail.

## Where to Park

Parking Location: N46.88414, W55.42019

This is a gravel parking area at the trail head.

## Walking Directions

Follow the hiking trail toward Chambers Cove for about 1 kilometre. Where the trail nears the shore east of Chambers Point, an interpretive panel explains the *Truxtun* disaster and rescue. There, instead of continuing west to Chambers Cove, walk east along the shore for about 100 metres. At a convenient spot, cross down onto the outcrops along the beach.

## Notes

For more information about the 1942 USS *Truxtun* disaster in nearby Chambers Cove, visit www.mun.ca/mha/polluxtruxtun.

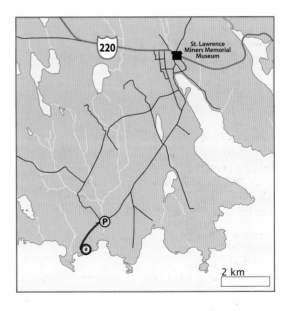

---

**1:50,000 Map**

St. Bride's 001L16

**Provincial Scenic Route**

Heritage Run

# On the Outcrop

Angular coastal outcrops reveal the fine, even texture of the St. Lawrence granite east of Chambers Point.

Outcrop Location: N46.87745, W55.42676

The most noticeable feature of the shoreline here is its colour. Shades of pink tint granite outcrops and beach shingle alike. If you look closely at the outcrops, you'll see mineral grains ranging in colour from rose to orange to brick red. Those are all grains of potassium feldspar, the most common mineral in the rock. The light grey grains are quartz. These two minerals make up more than 95 per cent of the rock.

Because the granite is low in aluminum, it does not contain any dark biotite or other mica (mica is aluminum-rich). Instead, most of the sparsely scattered dark grains in the rock are a type of amphibole called riebeckite.

You'll notice the outcrop is very blocky and angular. As the granite crystallized and cooled (and therefore shrank a little), three sets of cracks formed, all at roughly right angles (approximately north-south, east-west, and horizontal). As it weathers, the granite breaks apart along these directions, creating the blocky pattern.

| 1000 | 900 | 800 | 700 | 600 | 500 |
|------|-----|-----|-----|-----|-----|
| $Z_1$ | | $Z_2$ | | $Z_3$ | € |

# FYI

- St. Lawrence granite intrudes rocks formed on an ancient ocean floor about 760 million years ago, during the Cryogenian period (see site 35). That places the oldest rocks in the Avalon zone adjacent to one of the youngest.

- The chemical composition of the granite shows that it crystallized high in the crust, at a shallow depth. Since then, the terrain has not been eroded very much – visible outcrops are probably located near the top of the intrusion. In fact, some of the older sedimentary and volcanic rocks in the area may form the "roof" of the intrusion, with the granite lying just underneath.

- About 40 veins of fluorspar (also known as fluorite) occur within the St. Lawrence granite or related dykes. They formed in cracks that developed while the granite was cooling. The veins are several metres wide and up to 7 kilometres long. Where they cross into the surrounding rock, they quickly narrow down and disappear.

## Related Outcrops

The St. Lawrence granite is one of several Devonian granites (shaded green in the map below) in Newfoundland; most of them were intruded along or near the Dover fault separating the Gander and Avalon zones (see site 34).

## Exploring Further

St. Lawrence Miner's Memorial Museum, Route 220, St. Lawrence. The museum has displays related to the history of fluorspar mining, and a workshop producing crafts of local fluorspar.

Martin, Wendy. *Once Upon a Mine: Story of Pre-Confederation Mines on the Island of Newfoundland.* Special Volume no. 26. Quebec: Canadian Institute of Mining and Metallurgy, 1983. (Chapter VI outlines the history of fluorspar mining in St. Lawrence.)

| 500 | | 400 | | 300 | | 200 | | 100 | | 0 |
|---|---|---|---|---|---|---|---|---|---|---|
| Є | O | S | D | C | P | T | J | K | Cz | |

# Glossary

**algal mat** A layer formed as sediment is trapped by and accumulates in a community of algae living on an existing rock surface, typically in shallow or intertidal water.

**allochthon** A rock sequence or formation that has been tectonically displaced onto an adjacent terrain and thus did not form where it is now found. ah-LOCK-thon

**alluvial plain** A broad land surface created by an accumulation of river-deposited sediment, for example, filling a valley. ah-LOO-vee-al

**amphibole** A silicate mineral rich in iron and magnesium, often occurring as prismatic crystals in igneous or high-grade metamorphic rocks. Its colour depends on its chemical composition, but most commonly it is dark blue or green. AM-fib-bowl

**amphibolite** A metamorphic rock made primarily of the mineral amphibole and lacking quartz. Amphibolites can form by metamorphism of mafic intrusions or of certain sediments. am-FIB-o-lite

**andesite** A volcanic rock that is neither mafic nor granitic but intermediate between the two. AN-de-zite or AN-de-site

**angular unconformity** See **unconformity**.

**anorthosite** An igneous rock formed primarily from calcium-rich plagioclase feldspar, with small amounts of dark minerals. Most large bodies of anorthosite formed during the Proterozoic eon. ah-NORTH-o-zite

**Avalonia** A fragment separated from the continent Gondwana during the Ordovician period. It subsequently drifted as a separate microcontinent until its collision with Laurentia during the Silurian period. Named for the Avalon peninsula, where it was first recognized.

**basalt** An igneous rock formed primarily from plagioclase and pyroxene. Identical in composition to diabase and gabbro but with fine-grained crystals too small to see with the naked eye. bah-SALT

**basin** A depressed region of the Earth's surface surrounded by relatively high elevations. These include ocean basins but also basins formed by warping or faulting of continental crust.

**batholith** A complex igneous intrusion at least 100 square kilometres in area. Often granitic in composition, batholiths may include several distinct but related rock types formed by separate pulses of molten rock. BATH-o-lith

**bimodal intrusion** A body of igneous rock that contains only two distinct, contrasting rock types (for example, granite and gabbro) rather than a wide range of compositions.

**biotite** A dark brown or black mica that is rich in iron and magnesium.

**black smoker** A dark, chimney-like structure that forms on the sea floor where hot brine escapes due to volcanic activity. The brine contains sulphurous minerals that build up the chimney over time.

**brachiopod** Any member of a phylum of marine animals having two hinged but unequal shells. BRACK-ee-o-pod

**breccia** A rock containing a high proportion of angular, broken rock fragments cemented in a fine matrix. Breccias are named either for the process that created the fragments (for example, volcanic or tectonic breccia) or for the origin of the fragments (for example, limestone or pillow breccia). BRECH-ee-ah

**caldera** A crater-like depression created by the collapse of a magma chamber after it is emptied by a catastrophic volcanic eruption. call-DARE-ah

**Canadian Shield** A large region of gentle topography in eastern and central Canada and the north-central US that has remained geologically stable for at least 1,000 million years.

**cap carbonate** See **carbonate**.

**carbonate** A rock consisting of any combination of calcium and magnesium carbonate, including limestone and dolostone. A cap carbonate is a thin, persistent layer of such rock formed by rapid climate change following a global ice age.

**cephalopod** Any member of a class of marine mollusks that includes modern-day squid and nautilus. Many fossil species inhabited chambered, spiral-shaped shells. KEFF-ah-low-pod or SEFF-ah-low-pod

**chert** A form of silica ($SiO_2$) in which the individual crystals of quartz are too small to see (in some cases even with the aid of a microscope). It appears in a wide range of colours due to traces of iron, copper, or other elements.

**chlorite** A greenish silicate mineral with a scaly structure, somewhat similar to mica but easily scratched, commonly found in rocks affected by low-grade metamorphism. KLOR-ite

**concretion** A volume of sedimentary rock, often nodular, in which the mineral cement holding the sediment grains together is different from that in the rest of the rock. This can be due to biological activity or movement of fluids in the sediment while it is still soft and watery. con-CREE-shun

**conglomerate** A sedimentary rock containing a high proportion of rounded rock fragments larger than sand (pebbles, cobbles, or boulders) cemented in a finer matrix.

**conodont** The jaw-like fossil remains of an extinct class of marine animals possibly related to lamprey eels. CON-o-dont

**continental shelf** The nearshore portion of the continental margin, that is, the shallow marine environment between the shore and the continental slope.

**continental slope** The middle portion of the continental margin, that is, a region of steep slope and deepening water between the shallow continental shelf and deep continental rise.

**coticule** A metamorphosed, manganese-rich, deep marine sediment, possibly originating near hydrothermal vents. COT-ick-yule

**crinoid** Any member of a class of invertebrate marine animals with flower-like bodies attached to the sea bottom by a long stalk. Crinoids are related to starfish and share their five-fold symmetry. CRY-noid

**cross-bedding** Inclined deposits of sediment within a larger horizontal bed, formed along the edge of a ripple, bar, dune, or similar feature.

**debris flow** A landslide-like flow of material down a slope, resulting in a chaotic mixture of rock fragments scattered throughout a large volume of fine-grained matrix.

**delta** An often wedge-shaped accumulation of sediment at the mouth of a river, tidal channel, or submarine canyon, deposited as water flow slows down, causing sediment to settle out.

**diabase** An igneous rock formed primarily from plagioclase and pyroxene. Identical in composition to basalt and gabbro but with crystals just big enough to see with the naked eye. DYE-ah-baze

**dyke** A narrow, vertical or nearly vertical igneous intrusion formed when molten rock flows into a crack in an existing rock.

**dyke swarm** A cluster of dykes with similar age, composition, and orientation.

**extremophile** An organism that thrives in physically or chemically extreme environments that are deadly to most other life forms, such as extremes of temperature, acidity, or salinity. ex-TREEM-o-file

**fabric** The spatial arrangement of the mineral grains, rock fragments, or other constituents of a rock.

**fault** A plane or narrow zone along which a rock mass fractures and displacement occurs. Along a dextral fault or sinistral fault there is movement to the right or left respectively; along a thrust fault, one side moves over the other.

**feldspar** Any member of a group of silicate minerals in which silica, aluminum, and varying amounts of sodium, calcium, and potassium combine in a characteristic framework-like crystal lattice. Varieties include plagioclase and potassium feldspar.

**fluorspar** A rock-forming mineral, calcium fluoride, also known as fluorite. Impurities create a range of colour variations; some samples are fluorescent in ultraviolet light. FLOOR-spar

**foliation** A layered arrangement of mineral grains along which a rock easily splits, caused by the minerals' parallel alignment in response to compression by tectonic forces.

**foreland basin** A depressed region of continental crust, warped downward by an adjacent mass thrust onto the continent, for example, by collision or obduction.

**gabbro** An igneous rock formed primarily from plagioclase and pyroxene. Identical in composition to basalt and diabase but with larger crystals easily visible to the naked eye. GAB-roe

**garnet** A silicate mineral formed under conditions of high temperature and pressure, used as a gemstone and as an abrasive. Garnet is most often dark red, though green and blue varieties exist.

**gastropod** Any member of a class of invertebrate animals having a symmetrical, coiled shell that lacks internal chambers, such as snails. GAST-ro-pod

**gneiss** A coarse-grained, foliated rock with alternating bands of dark and light minerals. Gneiss can form from either sedimentary or igneous rocks under conditions of high-grade metamorphism. NICE

**Gondwana** A supercontinent of the geologic past that included areas found in the present-day continents of the southern hemisphere: India, Africa, South America, Australia, and Antarctica. gond-WAH-na

**granite** An igneous rock containing primarily quartz, plagioclase, and potassium feldspar, with or without small amounts of mica and/or amphibole. Granite is often cream, pink, or red in colour.

**granitic rock** Any igneous or metamorphic rock type dominated by a combination of quartz and feldspar. This is an informal field term applied to a variety of related rock types.

**graptolite** An extinct class of organism, fossil remains of which are preserved as clusters of black, carbonized impressions, often in shale. Abundant in the world's oceans during the early Paleozoic, they may have lived in floating colonies of interconnected tubes. GRAP-to-lite

**holdfast** A root-like structure that fastens marine organisms to rock or sediment on the sea floor.

**horst** A block of the Earth's crust that is uplifted relative to blocks on either side by vertical fault movements, often caused by rifting.

**hydrothermal** Part of, or relating to, a system of hot, circulating, often mineral-rich water, usually near volcanic activity.

**Iapetus ocean** An ocean of the geologic past that formed as the supercontinent Rodinia split apart, separating Laurentia from Gondwana. For much of its history, Iapetus lay in the southern hemisphere, near and roughly parallel to the equator. Its closure resulted in the Appalachian orogeny and led to the creation of a new supercontinent, Pangaea. ee-APP-ah-tus

**ichnology** The study of the physical traces left by animal behaviours such as travelling, feeding, and resting. Ichnology includes the study of both modern and fossil traces. ick-NAH-lah-gee

**intrusion** A body of molten rock that has travelled upward through the Earth's crust and invaded or displaced older rock, or the process by which this occurs.

**island arc** A curved chain of islands formed by volcanic activity above a subduction zone.

**kyanite** A metamorphic aluminum-silicate mineral that forms under conditions of high temperature and pressure. Named for the colour cyan, its crystals are usually blue blades or columns. KYE-ah-nite

**lamprophyre** An igneous rock containing high levels of potassium, in which biotite and/or hornblende often appear as larger crystals in a finer matrix that may contain feldspar. LAM-prah-fire or LAM-prah-fear

**Laurentia** A supercontinent of the geologic past that included areas found in the present-day continents of the northern hemisphere: North America, Europe, and Asia. lore-REN-chee-ah

**listvenite** (listwanite) A hard, often bright green rock characterized by a combination of quartz, chrome-rich mica, and calcite or other carbonate minerals, formed by the alteration of serpentinite. LIST-vah-nite

**mafic rock** Any igneous or metamorphic rock dominated by dark silicate minerals rich in magnesium and ferric iron, the two components from which the name is derived (<u>ma</u>gnesium, <u>fer</u>ric iron). MAY-fick

**magma** Molten rock.

**magma chamber** An underground reservoir of molten rock.

**mantle** The most voluminous layer of the Earth's interior, which lies between the iron-rich outer core and the silica-rich crust.

**matrix** The fine-grained portion of a rock in which larger crystals or rock fragments are embedded.

**melange** A rock formation large enough to be shown on a map, and consisting of a chaotic mixture of broken fragments in a finer, highly deformed matrix. The matrix is usually a weak, easily deformed rock type such as shale or serpentinite.

**metamorphism** The process by which the minerals in a rock recrystallize in response to changing conditions of temperature and pressure. met-ah-MOR-fizz-em

**mica** Any one of a group of aluminum-rich silicate minerals having crystals that form stacks or "books" of shiny, thin layers. MY-cah

**microcontinent** A fragment of crust wrenched or rifted from a larger, pre-existing continental mass.

**mid-ocean ridge** The long, narrow, mountainous area of an ocean basin along which new ocean crust is created by a continuous process of rifting and volcanic activity.

**Moho** A shortened name for the "Mohorovičić discontinuity," the boundary between the Earth's crust and mantle.

**mudstone** A sedimentary rock that originated as very fine sand and clay and that lacks the easily parted layering of shale.

**obduction** The tectonic emplacement of ocean crust and/or mantle onto continental crust along a convergent plate boundary. ob-DUCK-shun

**ophiolite** A sequence of rock types characteristic of spreading centres in ocean crust, including most or all of: deep ocean sediments, pillow basalts, sheeted dykes, gabbroic intrusions, and mantle peridotite. OH-fee-ah-lite

**orogeny** The process of mountain building. An orogen is the result of this process as preserved in the rock record, regardless of whether the region is still mountainous or has been worn down.

**overturned** A term applied to sedimentary sequences tilted more than 90 degrees from their original horizontal orientation, such that older beds are now above younger ones. This can occur when rock layers are folded.

**paleomagnetic** Pertaining to the alignment of magnetic minerals in a rock at the time the rock formed. Because magnetic minerals often align with the Earth's magnetic field, paleomagnetic data provide information about where on Earth the rock was located when it formed. PAY-lee-o-mag-NET-ick

**Pangaea** A supercontinent of the geologic past (approximately 300 to 175 million years ago). pan-GEE-ah

**peridotite** An ultramafic igneous rock consisting almost entirely of pyroxene and olivine. Peridotite is the most common rock type in the Earth's mantle. per-IDD-o-tite

**pillow lava** A form of basalt exhibiting bulbous shapes formed as lava erupts under water. Sometimes pillows shatter during this process, forming pillow breccias.

**plagioclase** A type of feldspar containing calcium and/or sodium rather than potassium. PLAJ-ee-o-klaze

**post-tectonic** Forming or occurring during a tectonically quiet period following an orogeny.

**potassium feldspar** A type of feldspar containing potassium rather than calcium and/or sodium. Potassium feldspar is often pinkish and lends granite its characteristic colour.

**pyrite** A metallic mineral, iron sulfide, sometimes known as fool's gold due to its shiny, yellow crystals. PIE-rite

**pyrophyllite** A aluminum-silicate mineral formed by severe hydrothermal alteration of granitic rocks and mined for a wide variety of industrial uses including the manufacture of ceramics and powder coatings. pie-ROFF-ah-lite

**pyroxene** A silicate mineral rich in iron and magnesium, similar to amphibole but with a simpler crystal structure that forms at higher temperatures. PEER-ox-een

**Rheic ocean** An ocean of the Paleozoic era that formed between Gondwana and several microcontinents including Avalonia. REE-ick or RAY-ick

**riebeckite** A dark-blue amphibole that is rich in sodium. REE-beck-ite

**rhyolite** A volcanic rock equivalent in composition to granite. RYE-o-lite

**Rodinia** A supercontinent of the geologic past (approximately 1,000 to 750 million years ago). roe-DIN-ee-ah

**schist** A strongly foliated rock rich in plate-like minerals such as mica. Schist can form from either sedimentary or igneous rocks during metamorphism. SHIST

**serpentinite** A greenish, waxy-looking rock formed by the alteration of peridotite, and containing one or more varieties of the mineral serpentine. ser-PENN-tin-ite

**shear zone** Similar to a fault in that both are caused by relative movement of adjacent blocks of crust, but in a shear zone, the movement is "smeared out" across a wide band instead of being focused along a single fault plane.

**spherulitic** A mineral texture characterized by small, rounded clusters of needle-like microscopic crystals, each cluster radiating from a common centre. SPHERE-you-LITT-ick

**striation** One of a set of usually parallel gouges or scratches on the surface of a rock, caused by glacial abrasion as rock-laden ice scrapes across a rocky landscape. stry-AY-shun

**stromatolite** A layered deposit of limestone built up over time as colonies of microorganisms (for example, blue-green algae) trap or precipitate the constituent minerals. stroh-MATT-ah-lite

**subduction** A process by which one tectonic plate moves under another along their common boundary. The setting where this occurs is called a subduction zone. sub-DUCK-shun

**supercontinent** A landmass that includes all or many regions of the Earth's continental crust, assembled by a series of continental collisions.

**syn-tectonic** Forming or occurring during a tectonically active period as part of an orogeny. SIN-teck-TAH-nick

**syncline** A fold in which the rock layers are tilted upward on either side of a midpoint, leaving younger layers nearest the centre; a downward fold. SIN-cline

**thrombolite** A clot-like deposit of limestone created by colonies of microorganisms that probably secreted or biologically aided the precipitation of constituent minerals. THROM-bo-lite

**thrust fault** See **fault**.

**thrust slice** A mappable sheet or slab of crust that was transported along a thrust fault into its current position.

**tonalite** An igneous rock containing primarily quartz and plagioclase, with small amounts of dark minerals such as biotite and amphibole. TONE-ah-lite

**trace fossil** Preserved evidence of animal activity, for example, in the form of tracks, burrows, feeding marks, or resting places.

**trilobite** Any member of a class of extinct marine arthropods (that is, related to insects, spiders, and crabs) having a distinctive central (axial) lobe flanked by right and left side (pleural) lobes. Trilobites evolved and became extinct during the Paleozoic era; more than 20,000 species have been identified. TRY-lah-bite

**trondhjemite** An igneous rock similar to tonalite but characterized by sodium-rich plagioclase. TRONN-jemm-ite

**turbidite** A sedimentary rock type deposited in deep ocean water by turbulent, avalanche-like currents of liquified sediment flowing down the continental slope. TUR-bah-dite

**ultramafic rock** A very dense igneous rock composed almost exclusively of dark silicate minerals rich in magnesium and ferric iron. Ultramafic rocks are predominant in the Earth's mantle but can also form as dense crystals accumulate at the base of a magma chamber in the Earth's crust. ul-trah-MAY-fick

**unconformity** An erosion surface preserved in a sequence of rock layers, representing a period of time during which no sediment was deposited. In an angular unconformity, sedimentary layers below the unconformity are tilted with respect to the younger layers, signifying an intervening cycle of deformation, uplift, and subsidence. un-con-FOR-mit-tee

**vent** A crack or pipe-like opening through which lava, steam, or other volcanic emissions reach the Earth's surface.

**vesicular** Containing empty cavities formed when bubbles are trapped in lava as it cools and hardens. veh-SICK-you-lar or vee-ZICK-you-lar

**volcanic arc** A linear or slightly curved pattern of volcanic activity above a subduction zone. Volcanic arcs located entirely within oceanic crust are called island arcs.

**xenolith** A rock fragment of contrasting appearance and origin occurring within an igneous intrusion. ZEEN-oh-lith

# Index of Place Names

# Image Credits

All photographs, maps, and diagrams in this book are by Martha Hickman Hild except as noted below.

Page 13. Adapted from Figure 14, "Early Ordovician to late Silurian tectonic evolution of the central Newfoundland Appalachians," by A. Zagorevsky, C.R. van Staal, and V.J. McNicoll, *Canadian Journal of Earth Sciences* (2007), vol. 44, p. 1582.

Page 15. Adapted from Table 1, "Stages in the Life Cycle of Ocean Basins and their Properties," by J. Tuzo Wilson, *Proceedings of the American Philosophical Society* (1968), vol. 112, no. 5, p. 312.

Pages 16 and 155. "Appalachian Orogen – Tectonic Lithofacies," postcard design by Dr. Clifford Wood, 1982.

Page 73. Adapted from Figure 14, "Schematic cross section through the Bay of Islands Complex," by H.S. Williams, *Field Trip A1: Geological Cross Section of the Appalachian Orogen* (GAC-MAC St. John's, 2001), p. 48.

Page 81, upper left corner. Fossil tree reconstruction by Arden R. Bashforth, *Late Carboniferous (Bolsovian) Macroflora from the Barachois Group, Bay St. George Basin, Southwestern Newfoundland, Canada*. Palaeontographica Canadiana, no. 24. St. John's: Canadian Society of Petroleum Geologists and Geological Association of Canada, 2005.

Page 139. Adapted from Figure 2, "The North Atlantic region in Upper Paleozoic and Lower Mesozoic time," by J. Tuzo Wilson, *Nature* (1966), vol. 211, no. 5050, p. 677.

Road maps were produced using datasets obtained online under the GeoGratis licence agreement from Natural Resources Canada (geogratis.cgdi.gc.ca). Although every effort has been made to provide helpful directions, under the terms of the GeoGratis licence, no representation or warranty of any kind is made with respect to the accuracy, usefulness, novelty, validity, scope, completeness, or currency of the data on which the maps are based.

Geological maps are based on datasets obtained in June 2011 from the Newfoundland and Labrador GeoScience Atlas OnLine, published by the Newfoundland and Labrador Geological Survey (gis.geosurv.gov.nl.ca/resourceatlas). The Geological Survey assumes no liability with respect to public use of this data.

# About the Author

Dr. Martha Hickman Hild received her PhD in Earth Sciences from the University of Leeds, UK, and has been a member of the American Geophysical Union since 1981. Early in her career, she co-directed a research laboratory and lectured in geology. She has worked as an editor in both technical and educational publishing, and as an award-winning news researcher. Since coming to Newfoundland, Martha has worked at Memorial University in a variety of administrative and report-writing roles. Her fascination with geology began at the age of four. A wilful and eccentric child, by age seven she felt the need to see inside rocks. To stop her bashing them against the sidewalk, her father relented and gave her a hammer. Thus began her geological adventures, which have taken her to Europe, Africa, Greenland, and elsewhere.